MAN AND BEAST

THE SOUTHWESTERN & MEXICAN PHOTOGRAPHY SERIES

This series originates from the Wittliff Collections, an archive and creative center established at Texas State University in San Marcos to celebrate the cultural arts of the region.

David L. Coleman, Series Editor

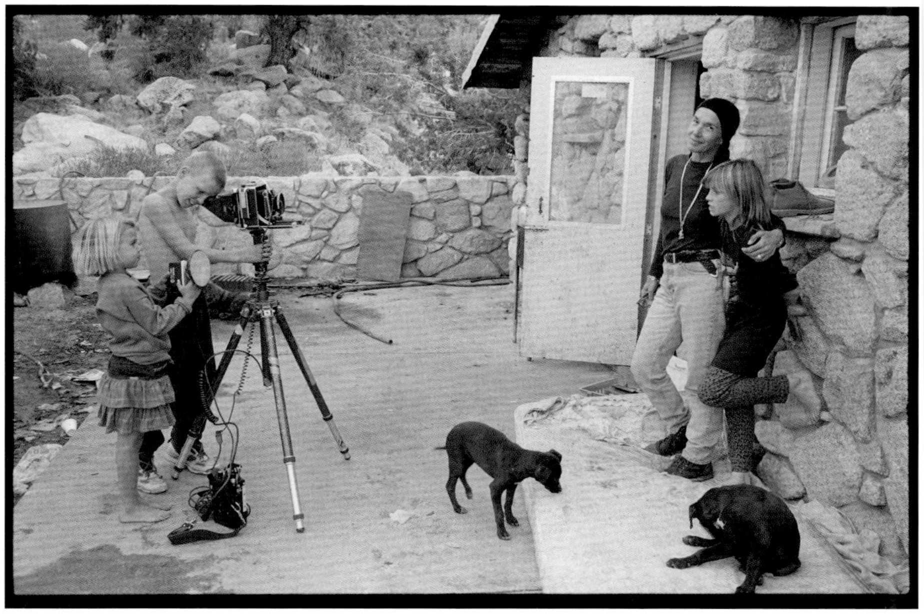

PHOTO BY GRANT DELIN

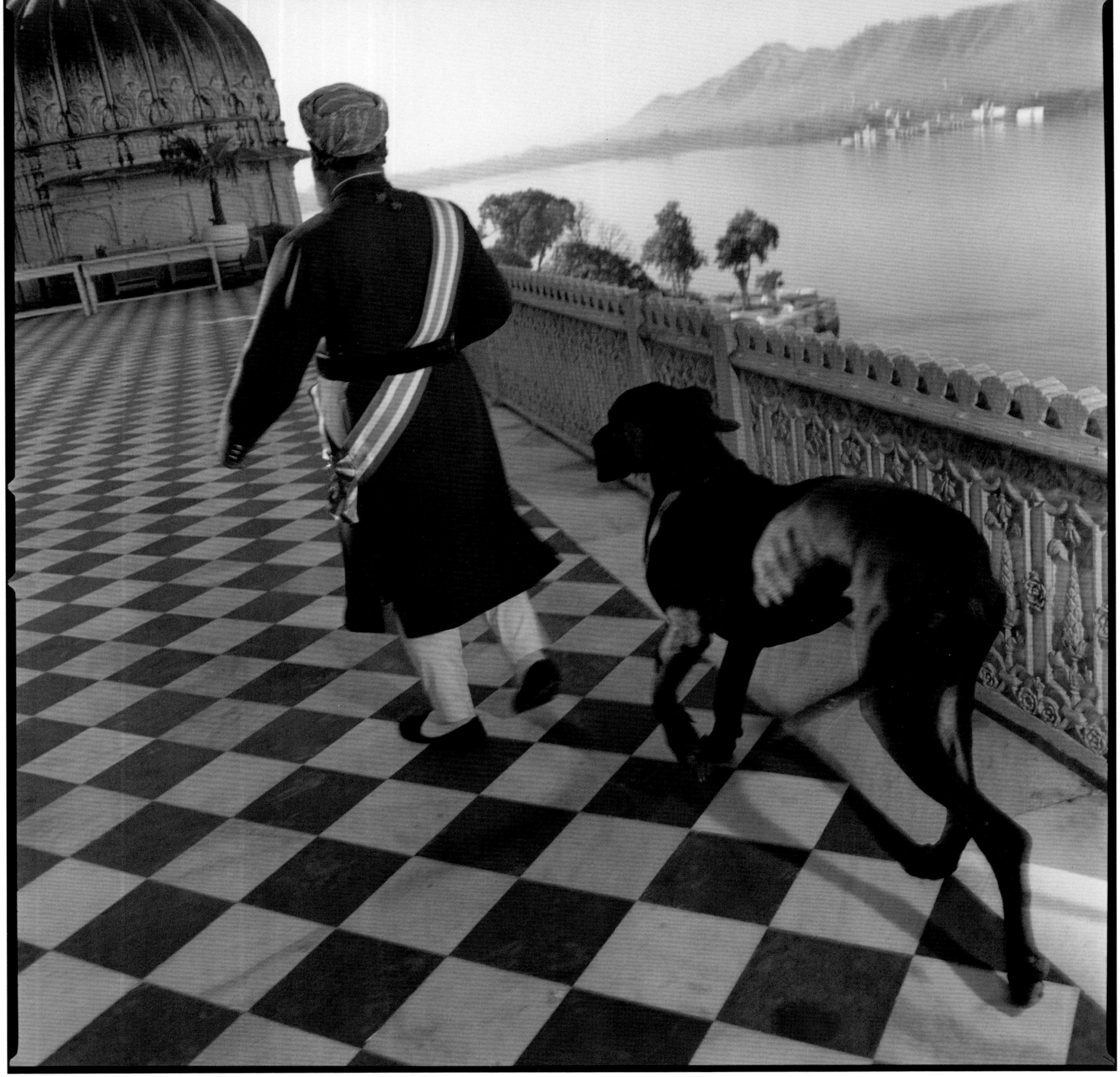

MAN AND BEAST

Photographs from Mexico and India

MARY ELLEN MARK

INTERVIEW BY MELISSA HARRIS

UNIVERSITY OF TEXAS PRESS, AUSTIN

Printed in China
First edition, 2014

Requests for permission to reproduce material from this work should be sent to:
Permissions
University of Texas Press
P.O. Box 7819
Austin, TX 78713-7819
http://utpress.utexas.edu/index.php/rp-form

♾ The paper used in this book meets the minimum requirements of
ANSI/NISO Z39.48-1992 (R1997) (Permanence of Paper).

LIBRARY OF CONGRESS CATALOGING-IN-PUBLICATION DATA
Mark, Mary Ellen, 1940–
Man and beast : photographs from Mexico and India / Mary Ellen Mark ; interview by Melissa Harris. — First edition.
pages cm. — (The Southwestern & Mexican photography series)
ISBN 978-0-292-75611-3 (cloth : alkaline paper)
1. Human-animal relationships—Mexico—Pictorial works. 2. Human-animal relationships—India—Pictorial works. 3. Animals—Mexico—Pictorial works. 4. Animals—India—Pictorial works. 5. Mexico—Social life and customs—Pictorial works. 6. India—Social life and customs—Pictorial works. 7. Mark, Mary Ellen, 1940—Interviews. 8. Human-animal relationships. 9. Animals (Philosophy). 10. Anthropomorphism.
I. Harris, Melissa. II. Title.
QL85.M3125 2014
591.5—dc23 2013030263
doi:10.7560/756113

This book is dedicated to all the wonderful animals that have passed through my life:

Agnes, Anarkali, Anna Mae, Angel, Beast, Bert, Billy, Bob, Bobbie, Bobbie Blue, BonBon, Boone, Boris, Boston, Buddy, Buster, Buster Ryan Thode, Charlie, Chollo I, Chollo II, Cindy Sherman (the beagle), Conchita, Coolidge, Delilah Rose, Digby, Dr. Elephant, Ella, Evatima, Fat Cat, Genki, Harriet, Harvey, Henry Hudson, Iko, Isarr, Kika, Kit, Kora, Longue Langue, Lorenzo, Louie, Louise, Martha, Mick, Mickey, Milosh, Mink, Mira, Molly, Moo, Nacha, Nick, Nicolasa, Olive, Ozu, Pancha, Papoose, Pee Pee, Percy, Poppy, Raja, Ripley, Rocky, Rosy, Ruby Tuesday, Sambu, Shadow, Sherman, Shiva, Shyama, Simon, Stella, Strider, T-Bone, Tex, Thomas, Truman, Violet, Wheaties, Wilma, Woofer, Zina, the adorable Cooper, and dear old Gringo.

And of course, as always, to my husband, Martin,
who is sometimes also as wonderful as a dog.

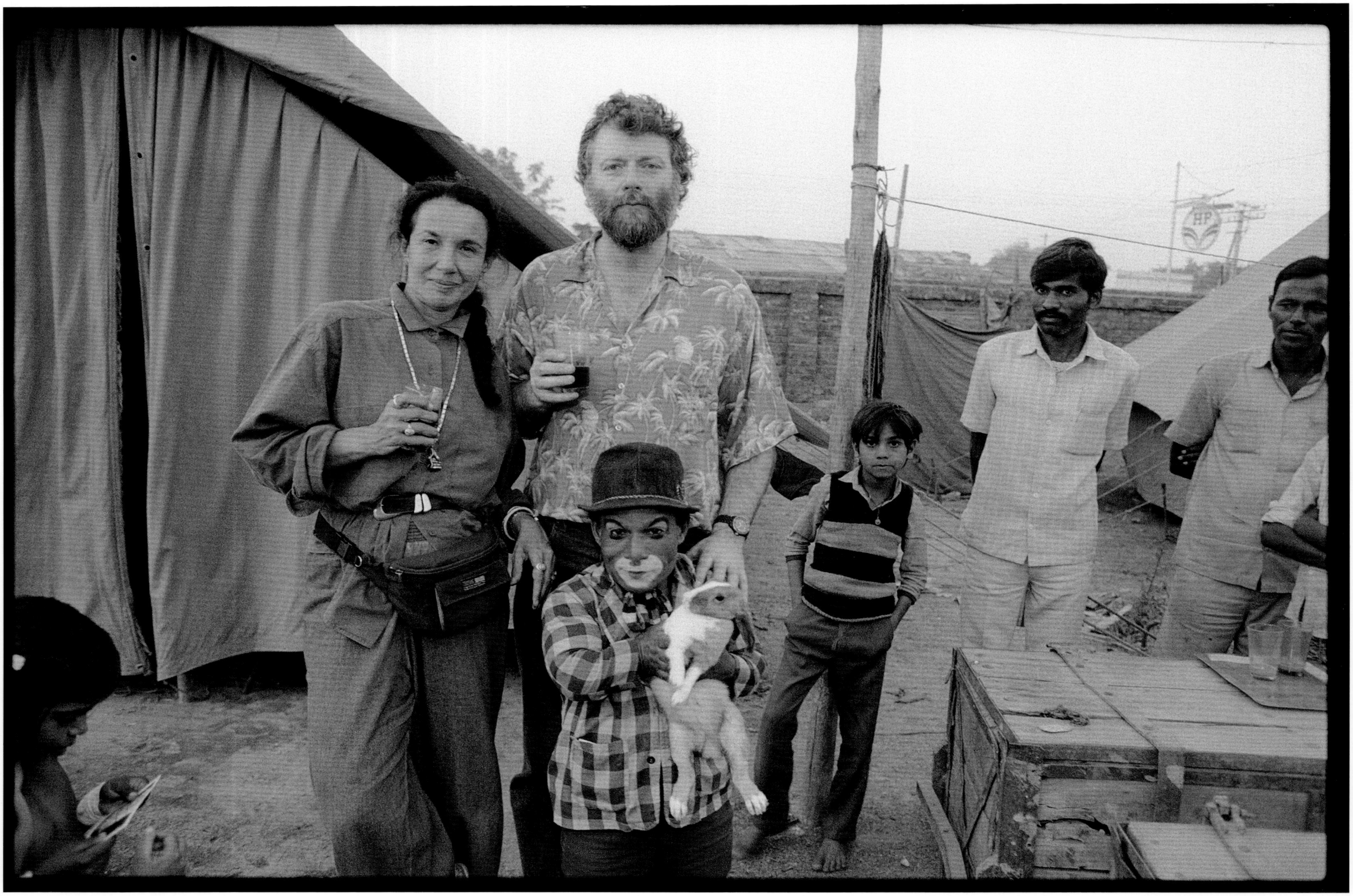

PHOTO BY DAYANITA SINGH

Photographer's Statement

Most of the images in this book are about the relationship between man and animals, the anthropomorphic quality of animals, and the animalistic quality of man. This was the perfect opportunity for me to use a title that has been in the back of my mind for years: *Man and Beast.*

My photographs often explore the balance between man and beast—whether it's literally a photo of a person with an animal or an underlying sense of the beast within human nature.

The photographs in this book are from two countries I love very much: India and Mexico. Both countries overwhelm my senses with their powerful imagery. They are very complex, intriguing, and inspiring. I've been traveling to both countries for many years. As strange as it may seem, I find strong similarities between these two cultures separated by thousands of miles.

There is a primal force I sense in the people and their culture that binds the two countries in my mind and it is this quality that makes the relationship between man and beast even more obvious. There is a more fundamental and intimate working relationship between the people and animals and this relationship is something I am drawn to and try to convey in many of my photographs.

The photographs in this book were taken as early as the late sixties and as recently as a few years ago. Some of them have been published before; most of them have not.

MARY ELLEN MARK

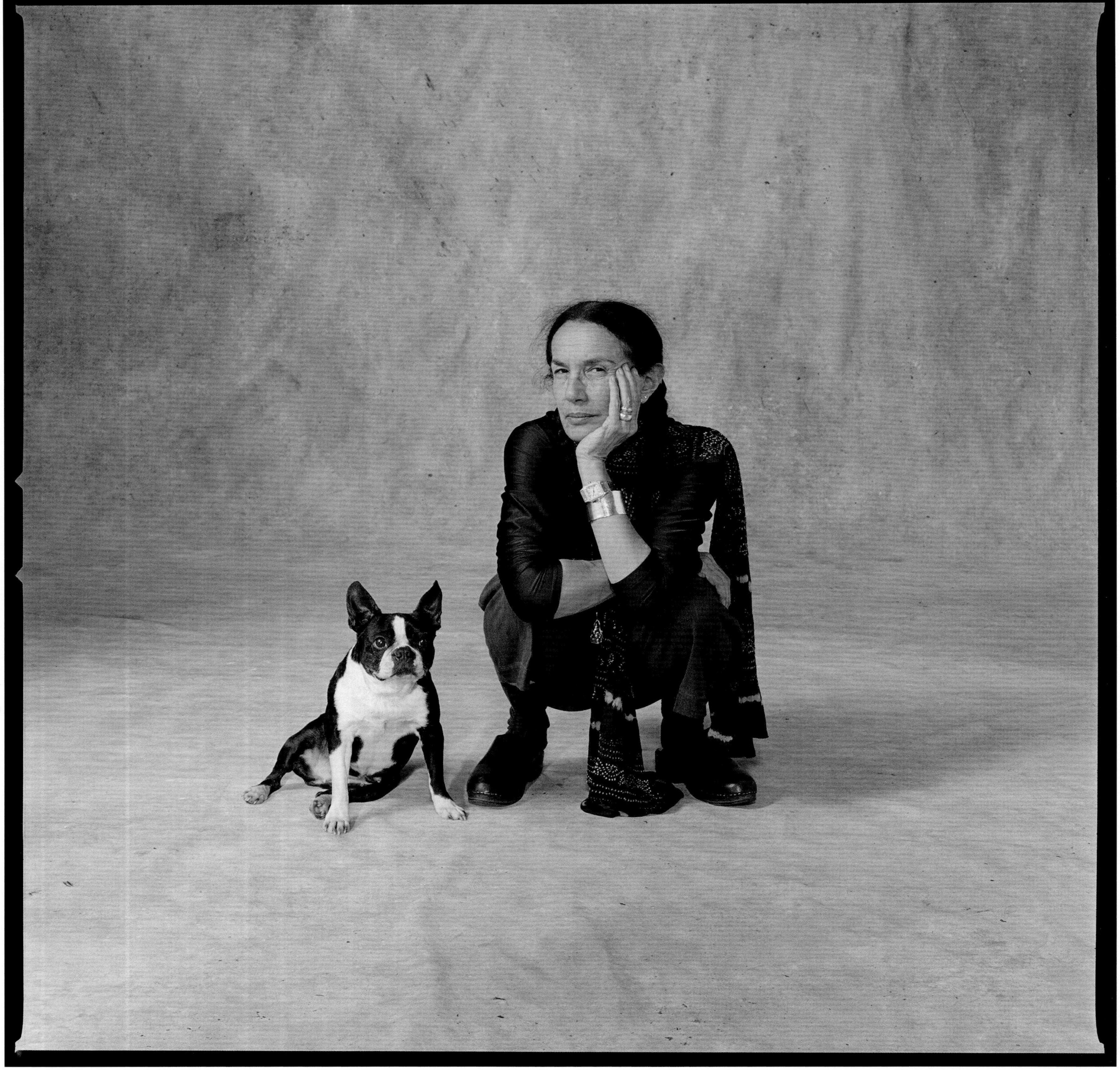

PHOTO BY MARK MOROSSE

MEM AND BEAST

In December 2002 I found myself walking across Washington Square Park with my then-one-year-old Lhasa Apso, Ella, who was (reluctantly) wearing a fuzzy, bubblegum-pink boa—part of a ballerina outfit for children (although Ella would have no part of the tutu—just too demeaning, given her roots as a Tibetan palace guard dog). It was during that walk that I began to fully comprehend Mary Ellen Mark's kooky, intense, and committed love affair with dogs. Ella and I were headed to Mary Ellen's annual "Doggy Christmas Party." Fifty to sixty dogs of friends and colleagues poured into her SoHo studio, both to attend the bash and to have their portraits taken. Sometimes at the party there are themes or props or backdrops or characters (an unfortunate soul dressed as the Statue of Liberty, for example, who posed with the dubious dogs)—but this year it was more or less simply about "festive attire," hence the boa. I placed Ella on the platform for the shoot, whispering sweet "Sits" in her ear, until the time came for me to sidle away, as Mary Ellen prepared to photograph her. But Mary Ellen was not yet convinced by the composition, so she called to Ella to move "stage right" and "upstage," along with other, non-doggy directions. Just an instant before Ella, overwhelmed, bolted decisively with cheetah-like agility, Mary Ellen made the portrait, which perfectly captures Ella's funny imperiousness.

Mary Ellen becomes deeply invested in many of her subjects—sometimes knowing and photographing them over the course of many years, as she did with Erin Charles (a.k.a. Tiny, of the 1980s project *Streetwise*), and with the Damms, a homeless family that Mary Ellen photographed many times. There is often a performative or interactive element to the photographs, whether they are of twins, prom-goers, or street children, whether she is on the streets of Oaxaca, Mexico, where she teaches workshops each year, or at a horse farm in Connecticut or a circus in Calcutta, or photographing dogs—pretty much everywhere.

Mary Ellen's sense of dogs' unconditional love is matched only by her belief that beasts are, unlike man, rarely if ever gratuitously cruel. This understanding infuses Mary Ellen's images with an unsentimental poignancy and a fully intentional anthropomorphism that, while sometimes ironic and other times unsettling, always render photographs that are remarkably engaging and winning.

MELISSA HARRIS   NEW YORK CITY   JANUARY 2013

MELISSA HARRIS: *Tell me about the dogs you've found homes for; you've been really mission-driven with this.*

MARY ELLEN MARK: Gringo was the first successful one.

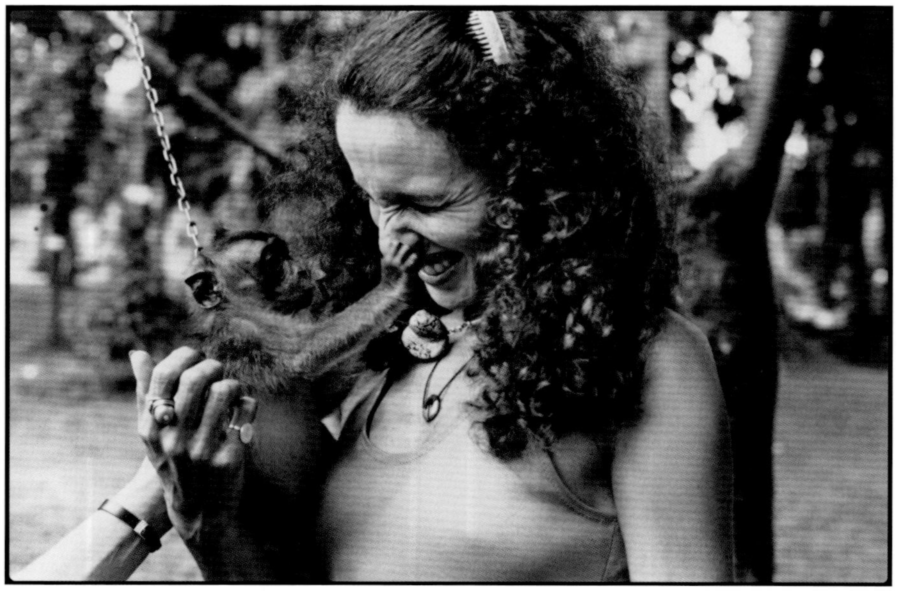

PHOTO BY EILEEN LOTTMAN

MH: *What happened with Gringo?*

MEM: Diana Haas [*art director*] took one of my early classes in Oaxaca and made pictures of him when he was a puppy. He lived with a family. She came back from taking pictures one day and said, "Mary Ellen, I met this puppy there, and I just love him—he's so adorable. But they torture him and throw hot soup on him." I was horrified. When that class was over, Diana left, but I stayed on in Oaxaca, as I was teaching a second workshop.

At some point she called me and said, "Oh, I miss that puppy!" I said, "Would you want it?" She said, "I'm going to raise it . . . please get that puppy." So I went to the family, and they said, "Okay, we'll get rid of this dog. But we want a poodle, and some money." So I had to give them money and we found a poodle for them. (I feel so guilty about this, because they didn't take good care of the poodle, either.) But anyway, it was a poor family. They were tough. They lived in the village. So we got Diana her puppy—Gringo.

And we brought Gringo back. He's a mutt. He loves Diana; he sleeps in bed with her. He has one fault. Because he was so abused, he does bite. He bit Meredith [*Lue, MEM's library manager*], and he's bitten me. But lovingly! He never forgets that I saved him. But he had a hard trip home, because the vet who organized everything sent him on a cheap flight, and he had to sit in Mexico City for a day in his little cage. We picked him up from the airport and took him out of his cage. He had never walked up steps before. We washed him off in the bathroom at the airport. I brought him a rawhide bone for the car. . . . When we got to Diana's place, I reached back to get the rawhide. Uh-oh—first bite!

The next morning Diana came over to our place, and Gringo was with her. He was sitting on her foot. He wouldn't go anywhere without her. It's been that way ever since.

Since then, I've taken it upon myself to find dogs for my Oaxaca workshop students, to convince my students to adopt dogs. And several of them do, and they have happy lives. I follow up.

There are a lot of stray dogs in Oaxaca, and the dogcatchers go after them. After I had a couple of fistfights with dogcatchers in the *zócalo*, they now let me take them. Certain dogs are just amazing and adorable, and I find homes for them.

MH: *And there was a hairless dog that you fell in love with . . . ?*

MEM: My friend Trine [*Ellitsgaard, a Danish weaver who lives with her husband, artist Francisco Toledo, in Oaxaca*] has a wonderful dog named Chollo. Chollo has had several litters since I've known her, and I've wanted to take a puppy home many times, but I haven't. . . . It wouldn't be a good home for a puppy. They are very sensitive, and I travel too much. It's a big responsibility; it's like having a child.

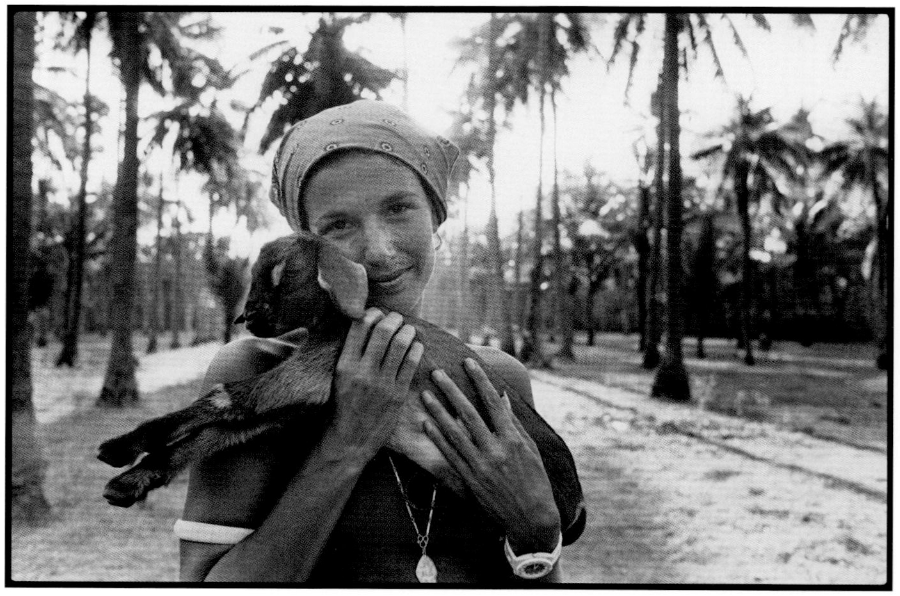

PHOTO BY JACK GAROFALO

But in Oaxaca, I would see a stray dog and fall in love with it, and then I wouldn't see it for a while. I'd wait and wait, and I'd go make sure it hadn't been picked up and given to the pound; I just couldn't stand going to the pound! One of my students photographed at the pound; the pictures are amazing. I couldn't even *go* there because it's so sad. But we do check to be sure that the dogs I know are not there . . .

When I was younger, I had horrible luck with pets. And I swore that as an adult I would make it up to them. I've never quite had the courage to own another animal myself. I've come close, but always, at the last moment, I realize I couldn't bear to cause harm to yet another dog or cat—or even a snake or hamster. Nonetheless, I love them and I love to photograph them, and I feel a great kinship to those who own and love animals.

MH: *What is your favorite animal?*

MEM: Well, I think I feel closest to dogs.

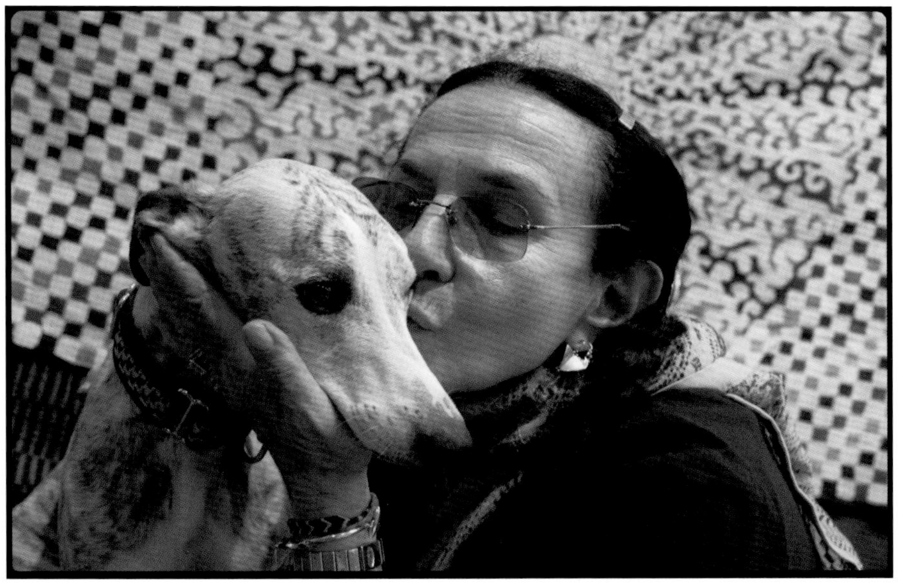

PHOTO BY MARTIN BELL

MH: *Is there a particular one?*

MEM: It wouldn't be fair to choose one—the others would kill me! I like all dogs for different reasons. I certainly love Julia's [*Bezgin, MEM's studio manager*] dog Cooper. He's a great dog; he comes here every day. And I'll always have a soft spot for Gringo.

Then there was the dog that I rescued from the Damm family.

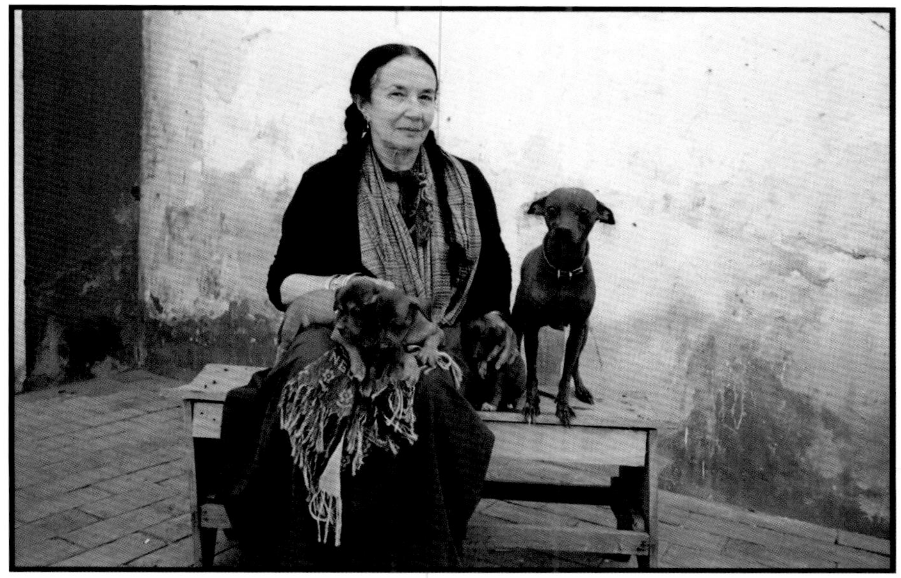

PHOTO BY MARCELA TABOADA

MH: *This was after you made the iconographic photograph of the Damm family living in their car, in the late 1980s, right? What happened with that dog?*

MEM: It was my second visit with the Damm family. I found them living in the high desert. They were squatting at an old, deserted ranch, and they had many dogs—maybe six. They were not caring for them at all. There was one that was fantastic, but the other dogs were really mean to him. He wasn't the master dog, so the alpha was mean to him. He was a big dog, and he was beautiful and so smart. He used to wait five miles down the road for us to arrive in our van and bring him food, and we did. The dog had a tag on his collar, so we were able to find out whom he belonged to. So I called and said, "Your dog is here. He's living with this family and they're not taking good care of him. I want to return him to you." They said, "We don't want him. We got rid of him." I said, "Well, don't you care? He's not being treated well." "No, we don't want him." So when I left, I decided I was going to take him with me and I'd find a home for him.

So we went out to the ranch, and my assistant and I took him. . . . We snuck him into the motel where we were staying, and we took him in the car. I don't think he had ever been in a car. He was terrified. He was shaking. We drove him to a friend's house in Los Angeles—it was about three and a half hours away. As soon as he saw my friend's house, he was like, "Bye, see ya!" She had a swimming pool and a lawn. He was right at home.

MH: *Your friend said yes?*

MEM: She took him. He immediately stole a chicken from her table—and she was having a dinner party that night! He was at home, living like a king—he was the happiest dog. Nick was his name. He was the most beautiful dog, and he had a great life. When I'd visit, it was clear that he always knew that I had saved him. Gringo knows that too. They are both smart; I like that. I like smart animals.

Recently I was photographing service animals. There was one really dumb dog. The first thing the trainer said to me when I got there to shoot the dog was, "We haven't quite mastered 'Sit' and 'Stay.'" I said, "Oh, great."

MH: *What was this dog doing in terms of service?*

MEM: It was like a petting dog, to make people feel good. But this dog did nothing. It was sweet. It was a pit bull. We took the dog to a school, but it was really hard to get a picture that wasn't sentimental.

Then we took photos of camels. We did two camels.

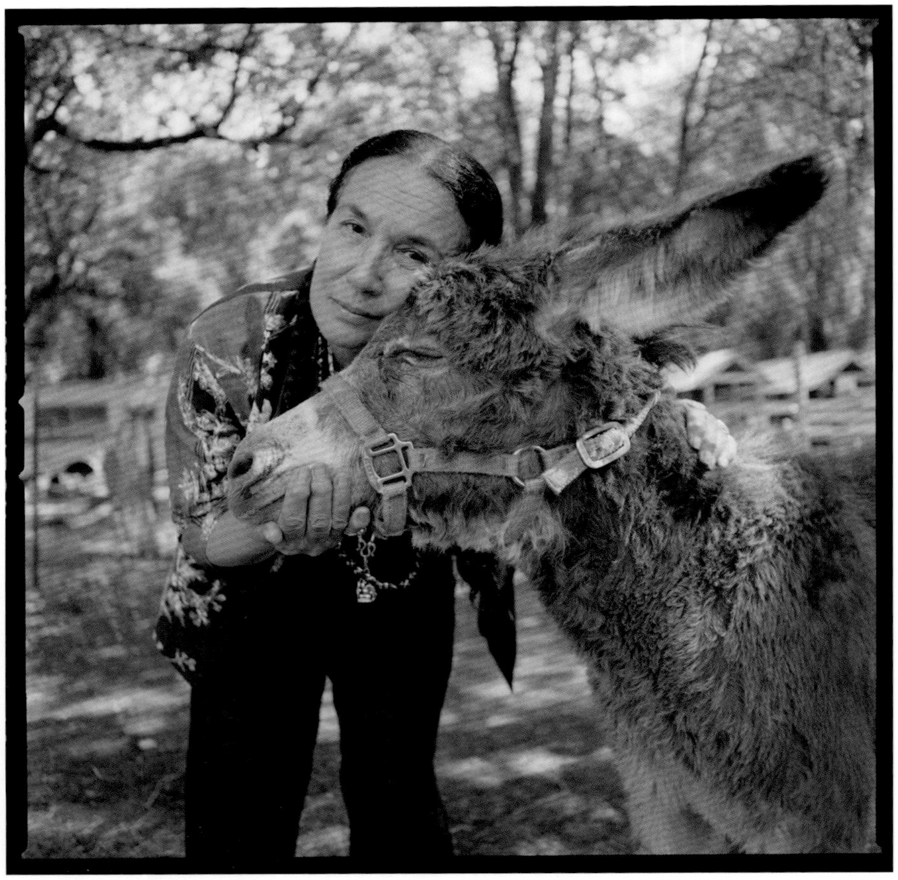

PHOTO BY MICHAEL PIAZZA

MH: *They're not supposed to be the nicest animals.*

MEM: They were babies, just a year old. They didn't do anything.

MH: *Where was this?*

MEM: This was in upstate New York.

MH: *What was their service role?*

MEM: They were at a school for children who have cognitive disorders—mild cognitive disorders, nothing serious. It's like a farm; it's a great place. And they have these camels there, and they get the kids working with them. So basically, the camels don't do anything except be there. They're gentle, so you can pet them. They were fine, they behaved. And they were strange looking, so that was interesting.

Then we photographed Billy, who was fantastic, at a barn in Connecticut, a riding school, mostly horses. Billy has one of the best trainers I've ever met; she adopts horses who have been badly treated, as well as old horses. But Billy is a donkey. Billy is just an amazing animal. This trainer loves him very much. Billy takes direction, in a way. You can hug him and he stands there. Donkeys are kind and smart. I think they're misunderstood.

Then we photographed a pig, who was an idiot! There is this place in New Jersey that's a home for people living with HIV. The pig goes and visits them. What did you think of the pig, Martin [*Bell, filmmaker and MEM's husband and sometimes collaborator, who is in the studio, within earshot of our conversation*]? That pig was an idiot, wasn't he?

MARTIN BELL: He had a very limited scope of operation.

MEM: I like animals you can relate to. And it's interesting to observe them in their roles with people.

One of the things that fascinates me most about the circus is the relationship between the performers and the trained animals—how they depend on each other for life and for work. I know the animal-rights people are going to hate this, and I understand there is a concern about how circus animals are treated, but I myself didn't see abuse—with only one exception. The animal trainers need the animals to survive, so they must treat them extremely well to make sure they are happy and in good health, and often there is a bond between them.

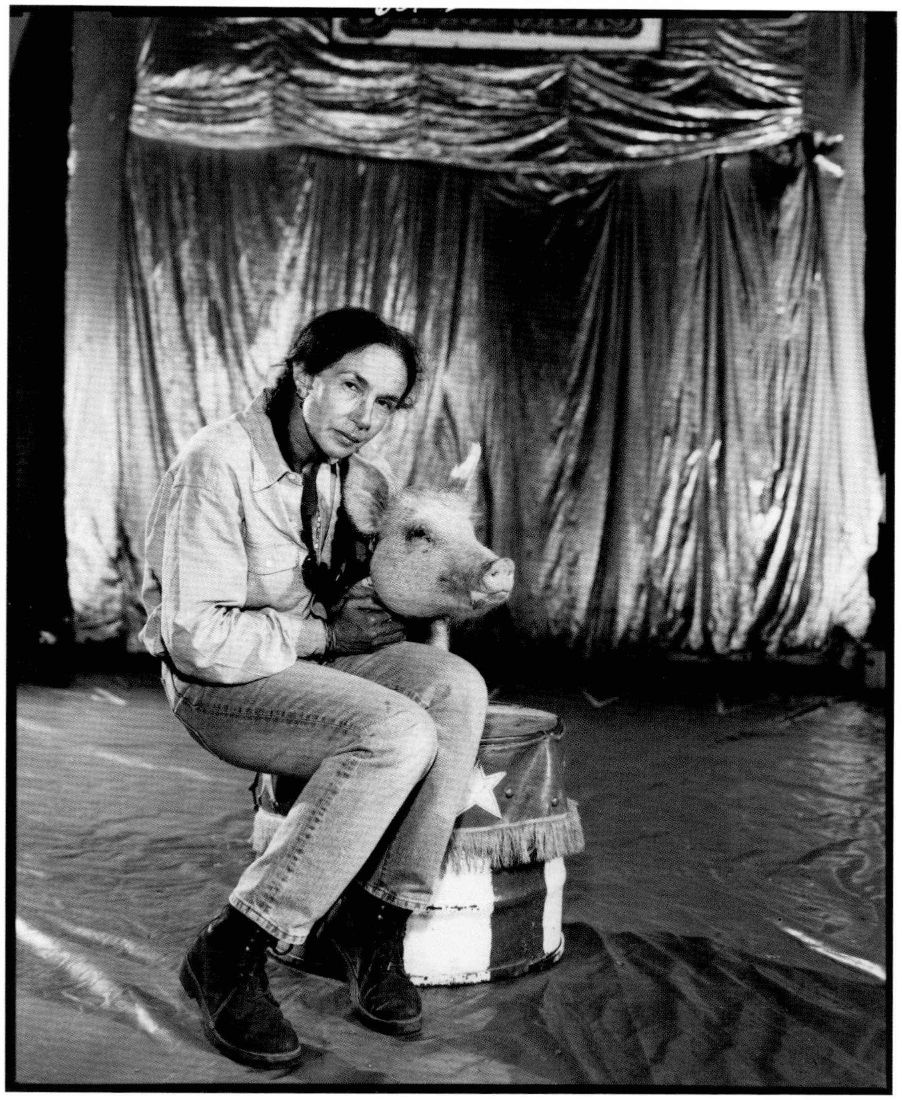

PHOTO BY JON NAIMAN

I also respect most animals, because they can kill you! I've never been attacked by an animal, but especially in the circus, I was careful. Although I did go in the cage with the lions in India. I felt I couldn't get a strong picture without being in the cage with them. Martin was in the cage, too. We were in the cage with twelve lions. That was frightening.

The trainer was in there, but he was a macho jerk. He smacked one lion—on the nose—to show off, which was a terrible thing to watch. The lion was so hurt that he jumped off the stool and started to run around the cage. We were just standing there, and Martin said, "Don't move!" The lion was bewildered and angry, and he was running around and running around—completely out of control. And there we were, just standing there. It was very scary. The trainer just wanted to show off, and he whipped him. It was horrible. We have the footage. We've never shown it because it is so disturbing. This guy was actually cruel—he was a cruel man. But he was unusual. Usually the trainers are not at all like that. . . . Mostly they depend on the animals. It's their livelihood; they have to treat the animal well. And, also, they love the animal.

MH: *Do they?*

MEM: Yes, they do love the animal. And they need the animal. They can't survive without it. And those animals are very expensive; they can't replace them. They have to be in good health. Trainers are very stupid if they treat the animal badly, because the animal can kill them, especially with an animal like an elephant. Elephants are very smart and sensitive. You have to be really respectful.

Chimpanzees bite. You can't trust chimps. One bit Martin. And, not only did it bite him—that chimp held a grudge against him! So the rest of the time we were shooting, the chimp looked for him. The trainer said, "I neglected to tell you that this chimp was trained in Germany, and he hates fair men. He *hates* them." Martin, what was the name of the chimp that bit you?

MB: Shiva.

MEM: So every time Shiva came out to perform, he was looking for Martin in the audience. We were there with [*the novelist*] John Irving this time, and every time the chimp would escape we'd say, "Where's John?" And then we'd find John, standing there with a chair in his hand!

MH: *So John Irving wasn't convinced about the chimp.*

MEM: No. That chimp was really dangerous. And another one bit me—a female chimp named Mira—I think because she was jealous. I shook hands with her trainer, and she was very jealous.

MH: *So she bit you?*

MEM: Well, I went to thank her trainer after I took her portrait with him; in the portrait his arm is around her. I shook his hand, and said, "Thank you so much." And the chimp looked at me and then she bit my hand.

The following year I came back to the same circus, and the trainer said to me, "Go and say hello to Mira. I'm sure she has forgiven you." So I went up to her cage—she was in a training cage, which was quite large. I put my arm in to shake her hand. She looked at me. Then she ran to the back of the cage and she took this stance. And I thought, "Oh my God, she's charging me!" I quickly pulled my arm out, just in time. . . . I never went around her again. She was really angry with me. She hadn't forgotten. They don't forget.

Another chimp I was very close to was named Raja. I loved Raja. Raja remembered me. I photographed Raja as a baby, and he remembered me when I came back. He loved me. I would go to his cage every day and give him a kiss, and he'd kiss me. The trainer said I was crazy, because he was huge—he looked like a gorilla. But he remembered me.

When I was leaving the circus, Raja knew I was going. He was so smart. The trainer said, "He's been a bad boy today; you can't go and say good-bye." But Raja started to cry, so I went anyway to say good-bye. He died a couple of weeks later. Chimps don't live long in captivity. Raja was only in his twenties.

Really, nothing compares with the Indian circus. I've thought about doing a major project on the Mexican circus—it's charming—but the Indian circus is much more strange and magical. The Mexican circus is too influenced by commercialism, whereas the Indian circus is so much more original. It reminds me of Fellini. Every day we were presented with all these incredible experiences.

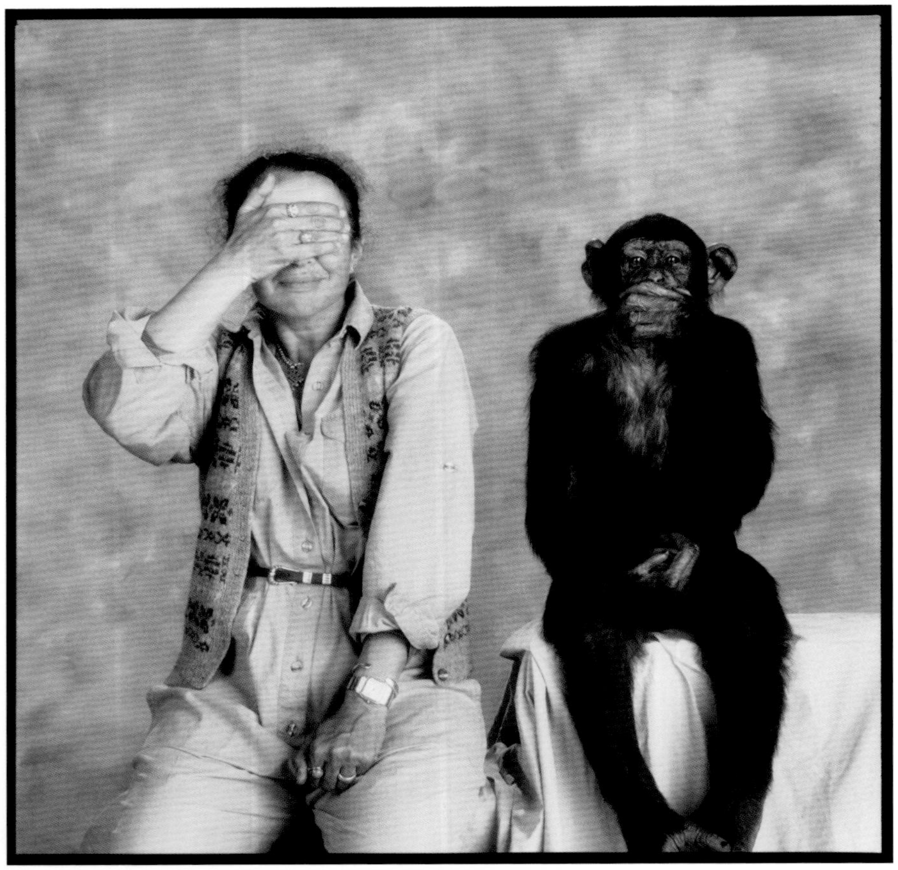

PHOTO BY FRANK SCHAEFER

MH: *Has Fellini been a real influence for you? I know that you photographed him on the set of* Fellini Satyricon *in 1969.*

MEM: I'd say yes. His movies are a real inspiration. I like the humor and sadness, pathos and magic. He's everything. His imagery is amazing. He's one of the most visual of directors. Bergman is also very visual, as is Tim Burton, whom I've worked with a lot—but Fellini is really otherworldly.

There are so many other great Italian directors. Vittorio De Sica is fantastic. There's that great film about the dog. Martin, what's the film about the dog? Is that De Sica? About the little dog?

MB: *Umberto D.* Flike is the dog's name.

MEM: *Umberto D.* Yes. A little dog. It's really beautiful.

MH: *Are there other photographers who focus on animals whose work is meaningful to you?*

MEM: Well, Nick's [*Michael "Nick" Nichols, natural history photographer*] work is wonderful. I love Nick because of his bravery and love of animals. I love the way he's able to integrate himself into the lives of animals in their habitats. He becomes part of them, which is extraordinary; no one else does that. He's not showing us anything sentimental about animals. He's a realist. So we're seeing the intimate side of the lives of animals. I'm amazed he can do that. He's patient. The pictures are really beautiful. They're interesting and beautiful. I love the fact that he's able to show us that intimacy.

MH: With Man and Beast, *what's the nature of Man these days?*

MEM: I wish that the nature of Man were better. I always think the nature of Beasts is better than the nature of Man. It's more pure, more honest. Beast is more honest than Man. Man is more deceitful, unfortunately.

MH: *When you take pictures of animals, what are you setting out to do?*

MEM: My goal is different from Nick's in the sense that I'm looking for an anthropomorphic side to them. Trying to show the "Man" in animals.

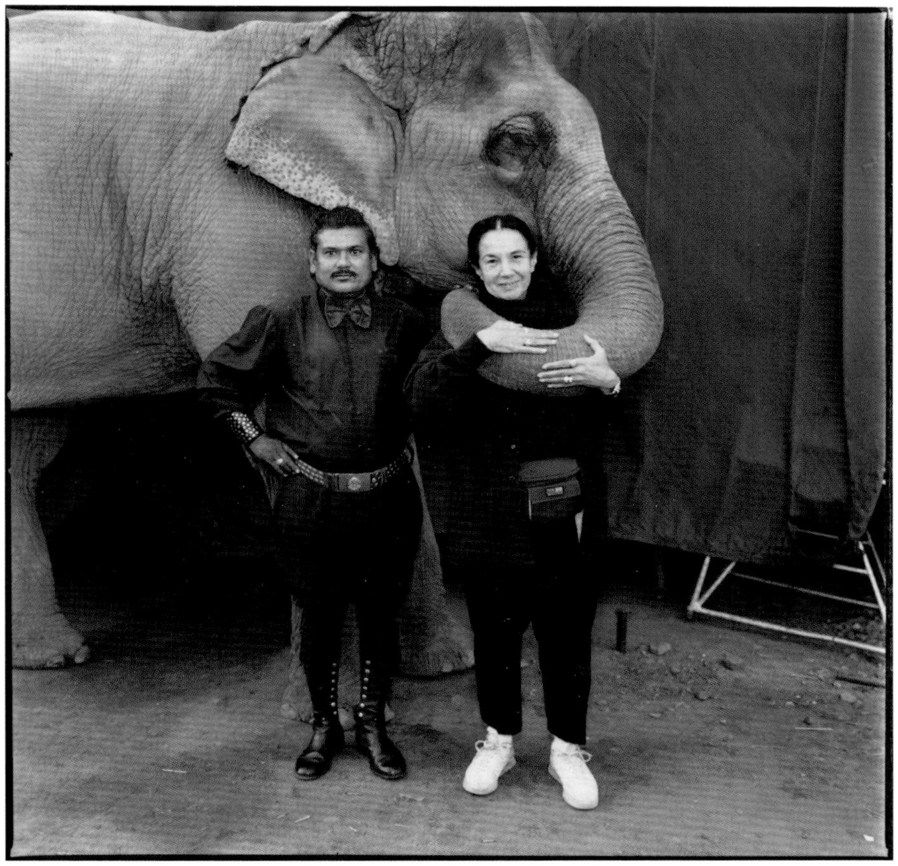

PHOTO BY FARROKH CHOTHIA

MH: *And what is that?*

MEM: Well, I'm trying to connect as a *human being* to what an animal is like. It's not like a human being at all, though; it's part of my fantasy. What really interested me so much in the Indian circus is that it's so much about the anthropomorphic side of animals, like "Dr. Elephant." Where else would you see a "Dr. Elephant"?

MH: *Well—the anthropomorphic side is what we determine, or read into their behavior, or force upon the animals. It's not really innate, is it? Continuing on the anthropomorphic track, is this ever demoralizing for the animal, from your perspective?*

MEM: I know I'm reading things into the animals' behavior, but on the other hand, their relationships, their roles with the humans in their lives are real—and sometimes, it's just so touching. . . . I photographed a female elephant who rides in the U.S.A.'s UniverSoul Circus. That elephant looks so sad to me—it's sad that it's trapped in the circus, that it's not out in the wild. I love elephants.

MH: *So you're exploring the emotional life of elephants?*

MEM: Well, again I'm reading that idea into it. . . . Maybe the elephant is sad because he wants a peanut! I don't know. It's just that you're looking, trying to read a certain emotion or narrative into it to get a great photograph. Like with the chimp Mira, when she had her arm around her trainer—I was trying to show that she was in love with him . . .

MH: *But we don't know what love means for a chimp.*

MEM: No, we don't. We do know that he feeds her.

MH: *Are children and animals similar to photograph?*

MEM: There's an innocence sometimes in children, which is similar in animals. But I rarely look for the innocence. Children are "Man." They can be very cruel. They're little people, after all. That's what I try to look for in children. I'm looking for the anthropomorphic things in animals. But I think in children I'm looking . . .

MH: *. . . for the Beast?*

MEM: Exactly. I look at children as just regular . . . as tiny humans. So I'm looking for the things they do that reveal their true nature.

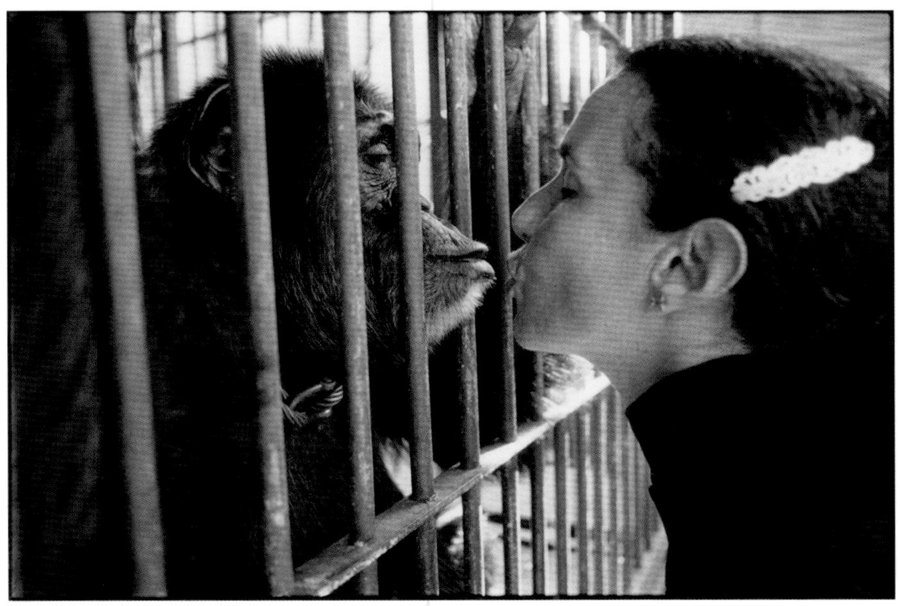

PHOTO BY DAVID LITTSCHWAGER

MH: *If you could be any animal, what would you be?*

MEM: Oh, gosh. Well, I don't know that I'd want to be a dog. . . . It would be nice to fly.

MH: *Would you want to be a hummingbird? Or a hawk?*

MEM: Not a hawk. I don't want to swoop down on little animals.

MH: *A seagull?*

MEM: What would I want to be? Maybe a very loved and spoiled dog. Or . . . a bird. A very free bird. Martin, what's a very *free* bird?

MB: What?

MH: *I asked Mary Ellen if she could be any animal, what she would want to be—*

MEM: Certainly not a penguin. I hate cold weather.

MH: *—and she only wants to be a dog if she's very loved and spoiled.*

MEM: A spoiled and loved dog, but owned by some very rich person who traveled, who took me all over the world.

MH: *She doesn't want to be a hawk, because she doesn't want to go after little animals. I brought up possibly a hummingbird . . . but that didn't seem to interest her.*

MB: Too small.

MH: *So what's the right animal for Mary Ellen to be? Apart from a spoiled dog?*

MEM: I wouldn't want to be an elephant. Too big.

MH: *Well, if you were an elephant not in a zoo, but in its natural habitat . . .*

MB: Or an elephant with Alan [*Root, naturalist and wildlife filmmaker*]. It was so nice when we were visiting him.

MEM: Those elephants were fat. [*She asks Martin:*] A giraffe?

MB: Why are you asking me?

MEM: No one has ever asked me this question before.

MB: You're entering new territory here.

MH: *We've been talking about anthropomorphism, so I'm turning it around now on Mary Ellen.*

MB: Yes. If you could be anything, what *would* you be?

MEM: Maybe a horse. That would be wonderful. A wild horse, living free . . . Martin, what's a wild horse?

MB: A mustang.

MH: *Do you love horses?*

MEM: I love images of horses.

MH: *But you don't really photograph horses.*

MEM: I don't photograph them. I don't ride. But they are beautiful animals.

MH: *So, a mustang.*

MEM: Yes, a wild horse.

MH: *Are they smart?*

MEM: Not very.

MH: *But you like smart animals.*

MEM: Yes, but horses are so beautiful. I certainly wouldn't want to be Billy the donkey. Although I loved him. Maybe a stallion. An Arabian. I'd like to be an Arabian stallion.

MH: *Martin—we've settled on an Arabian stallion.*

MB: Excellent news.

THE PLATES

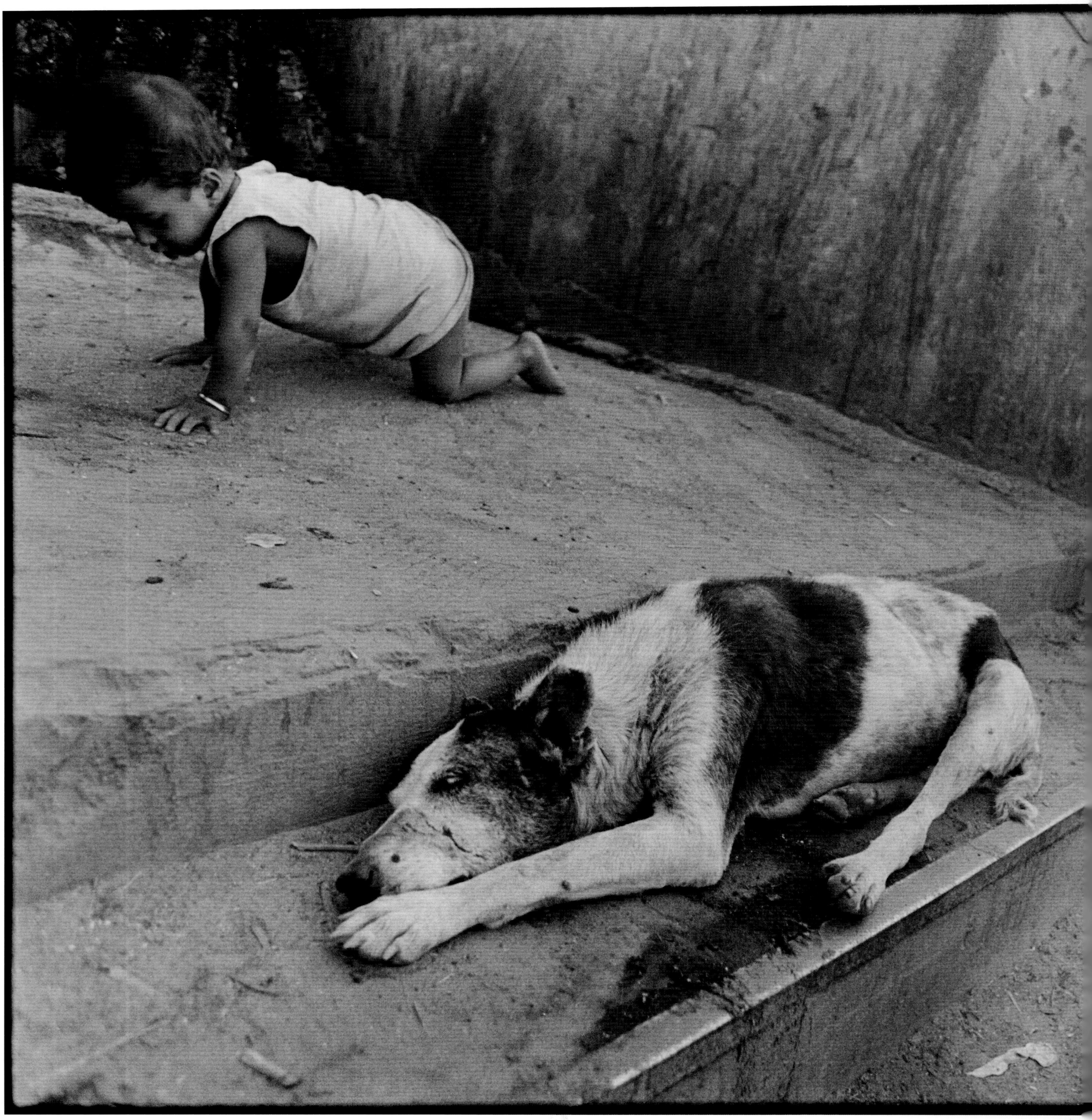

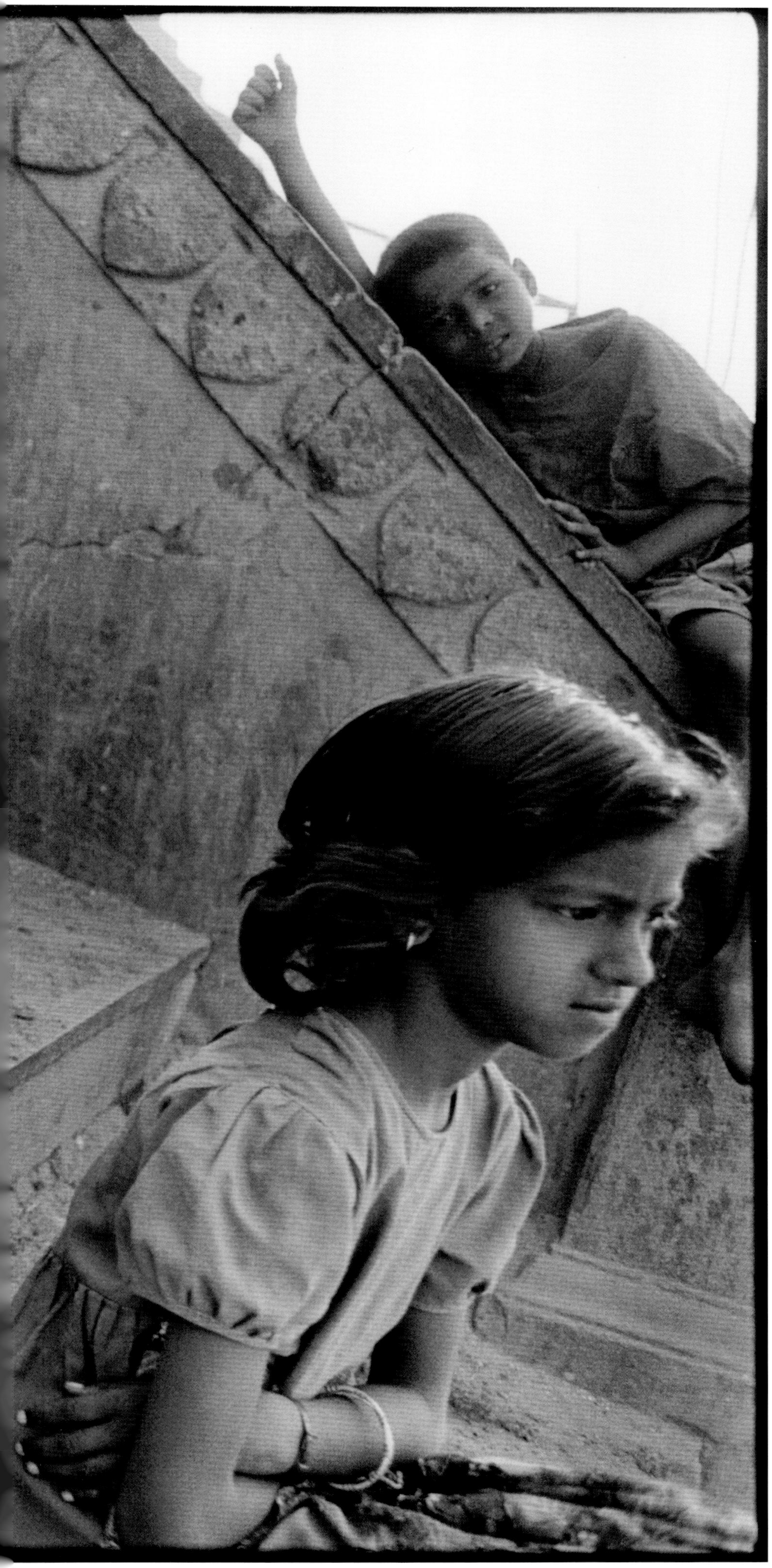

PLATE 1

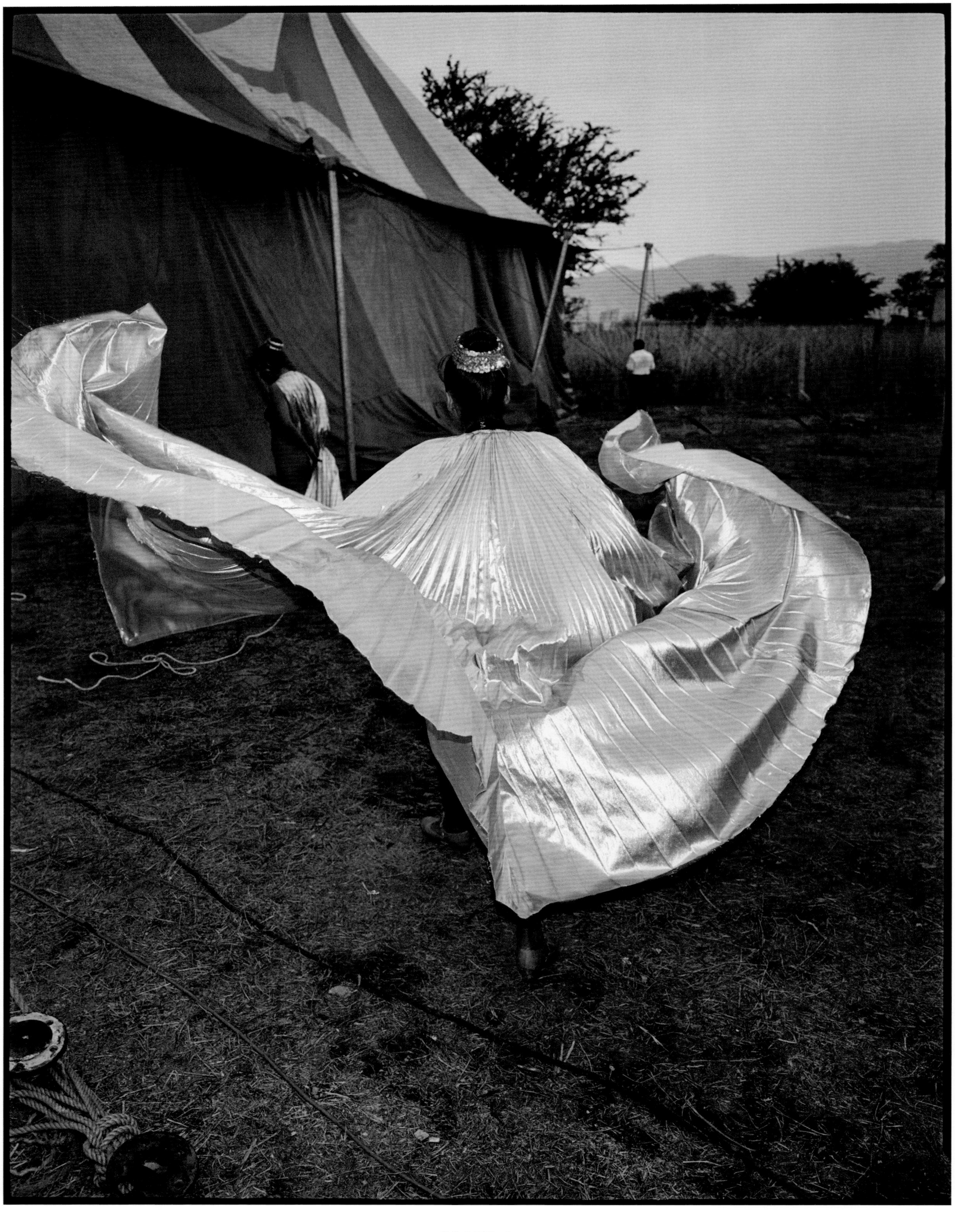

PLATE 2

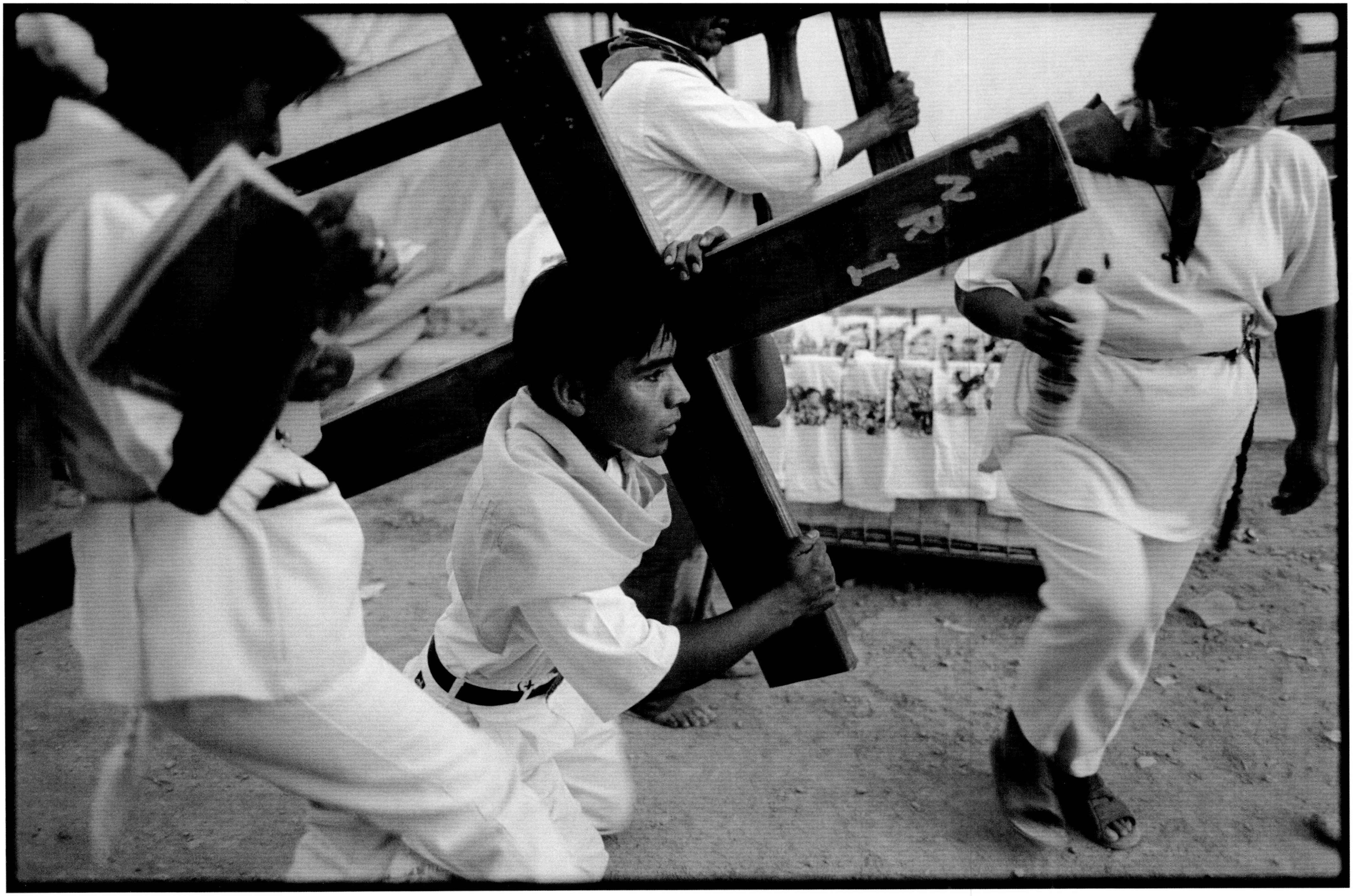

PLATE 3

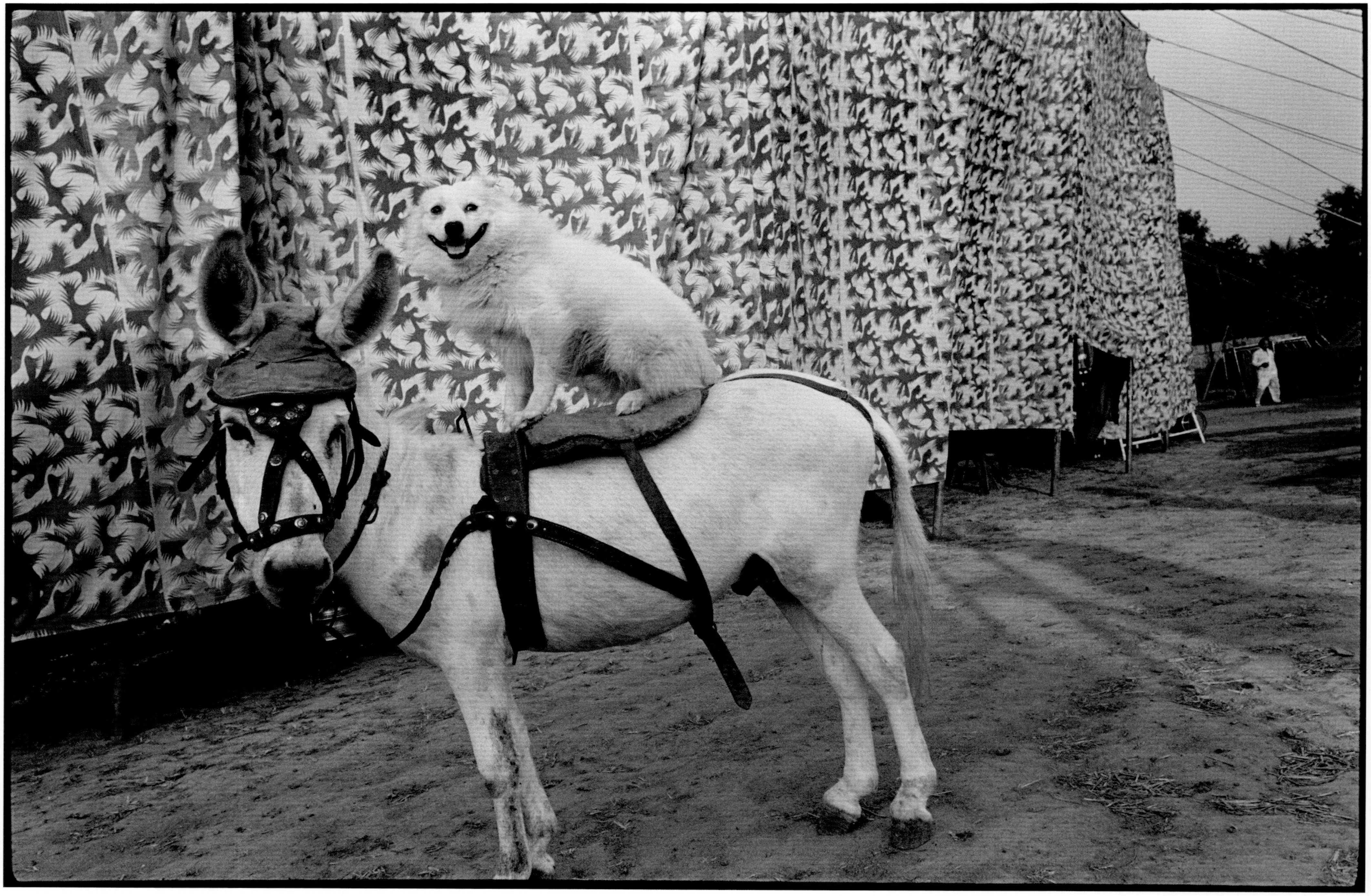

PLATE 4

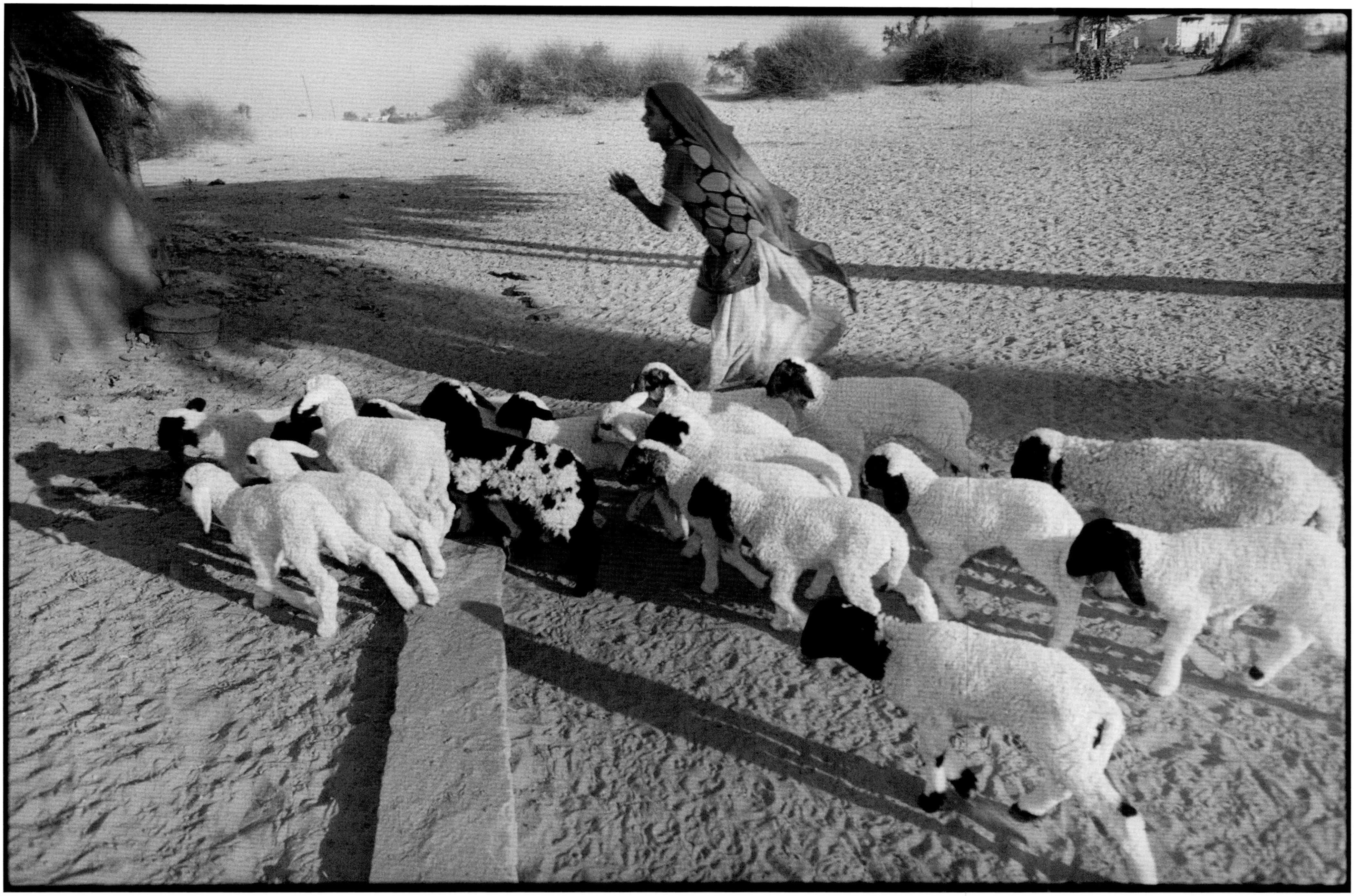

PLATE 5

PLATE 6

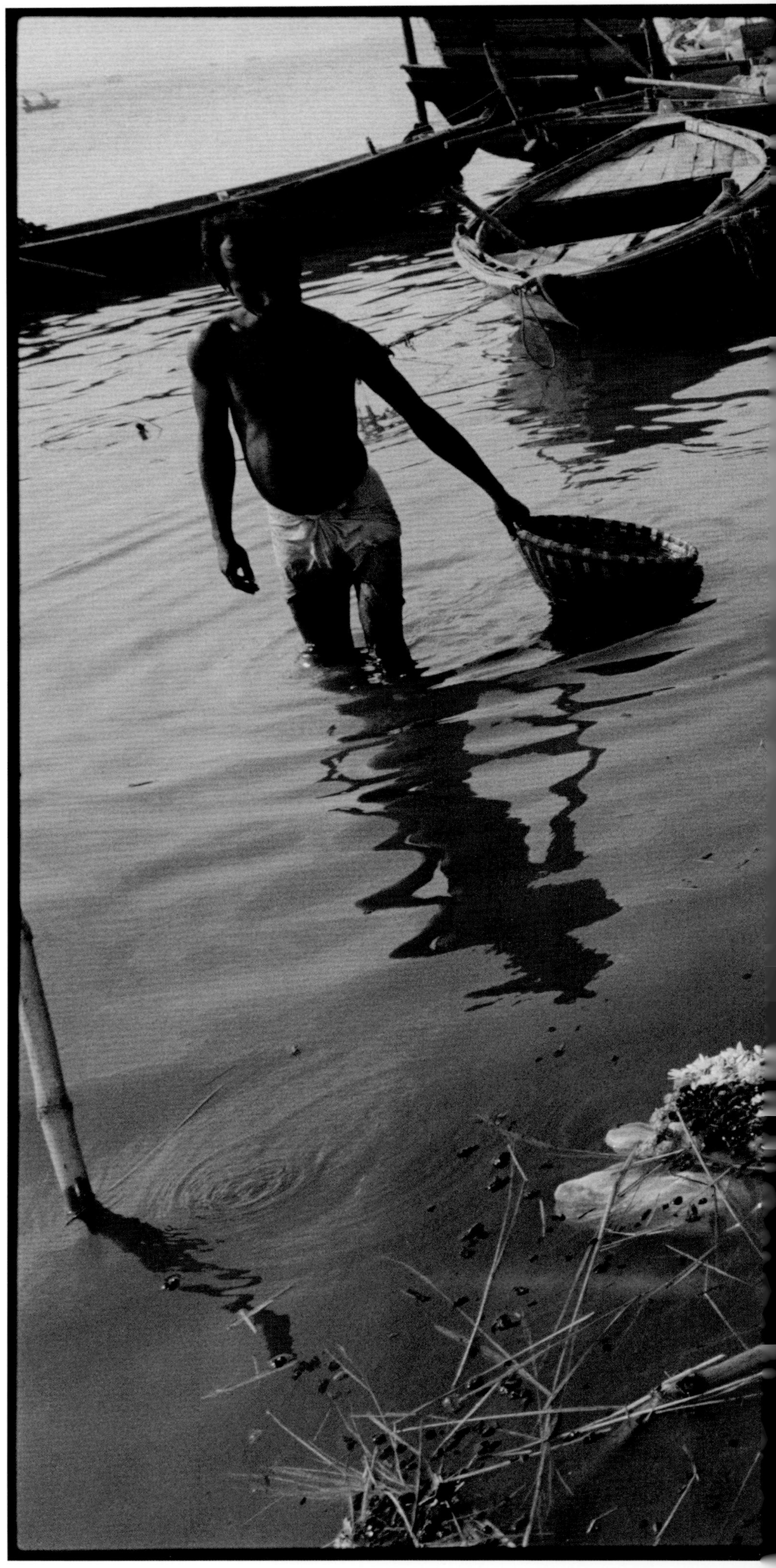

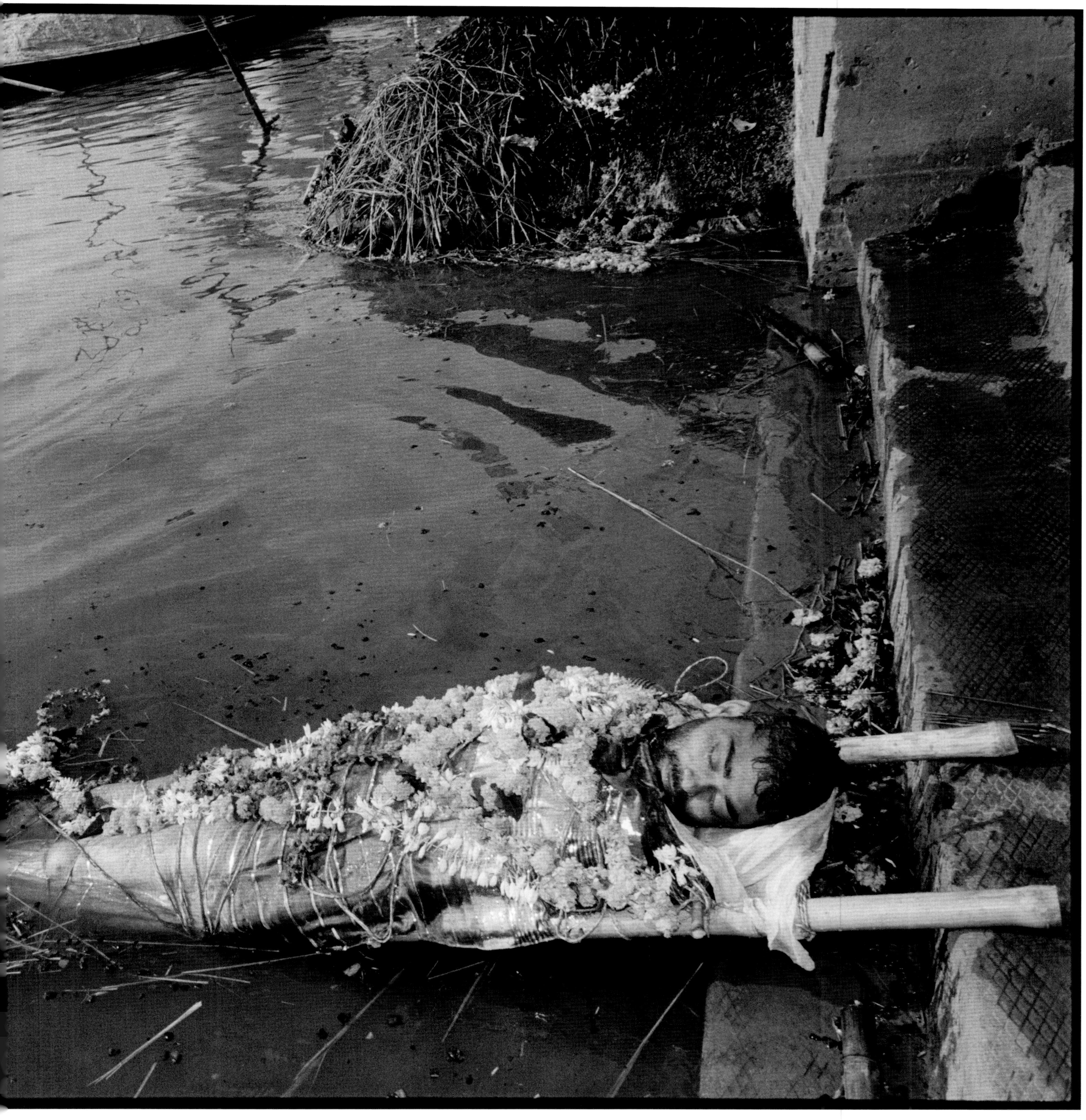

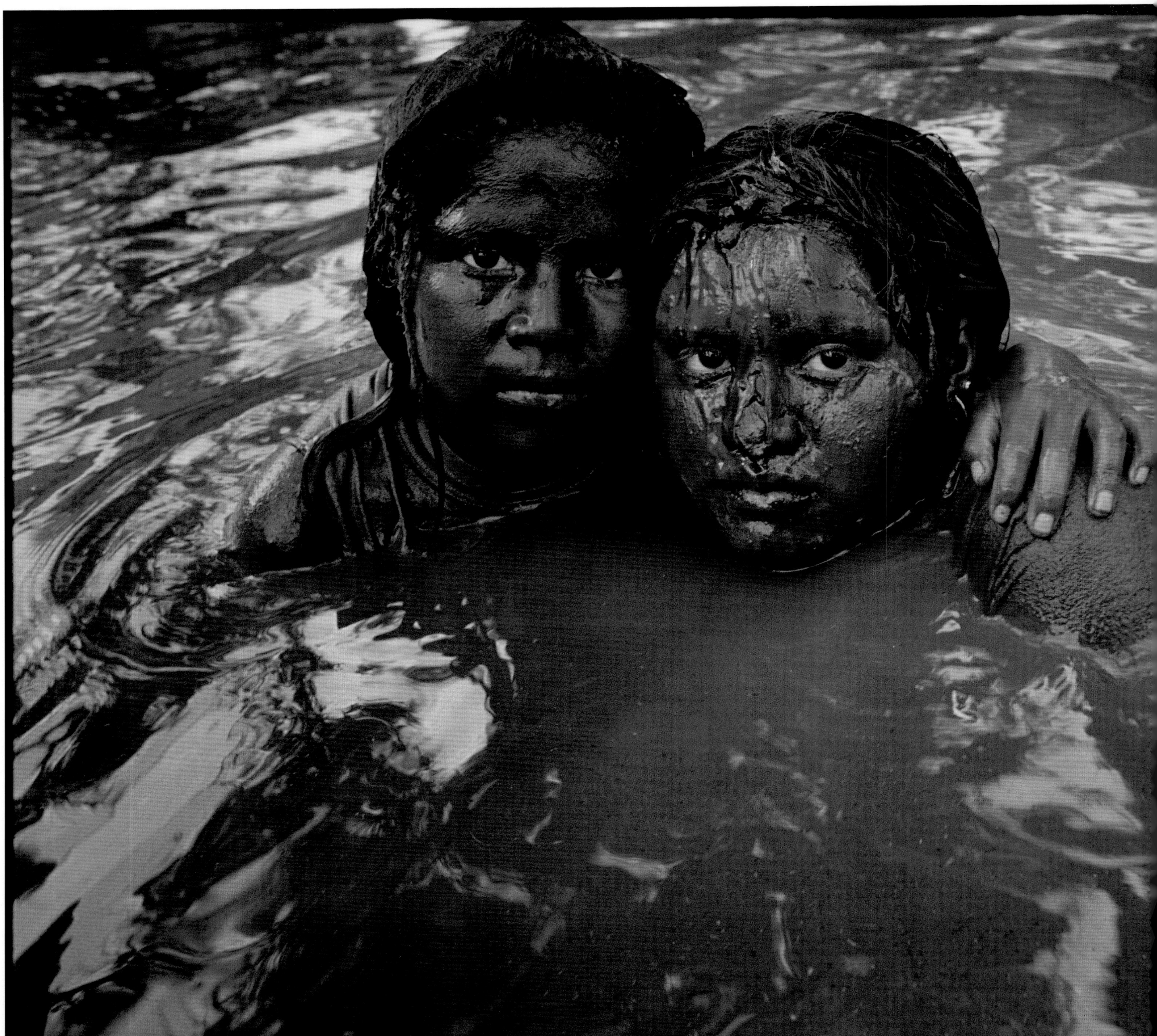

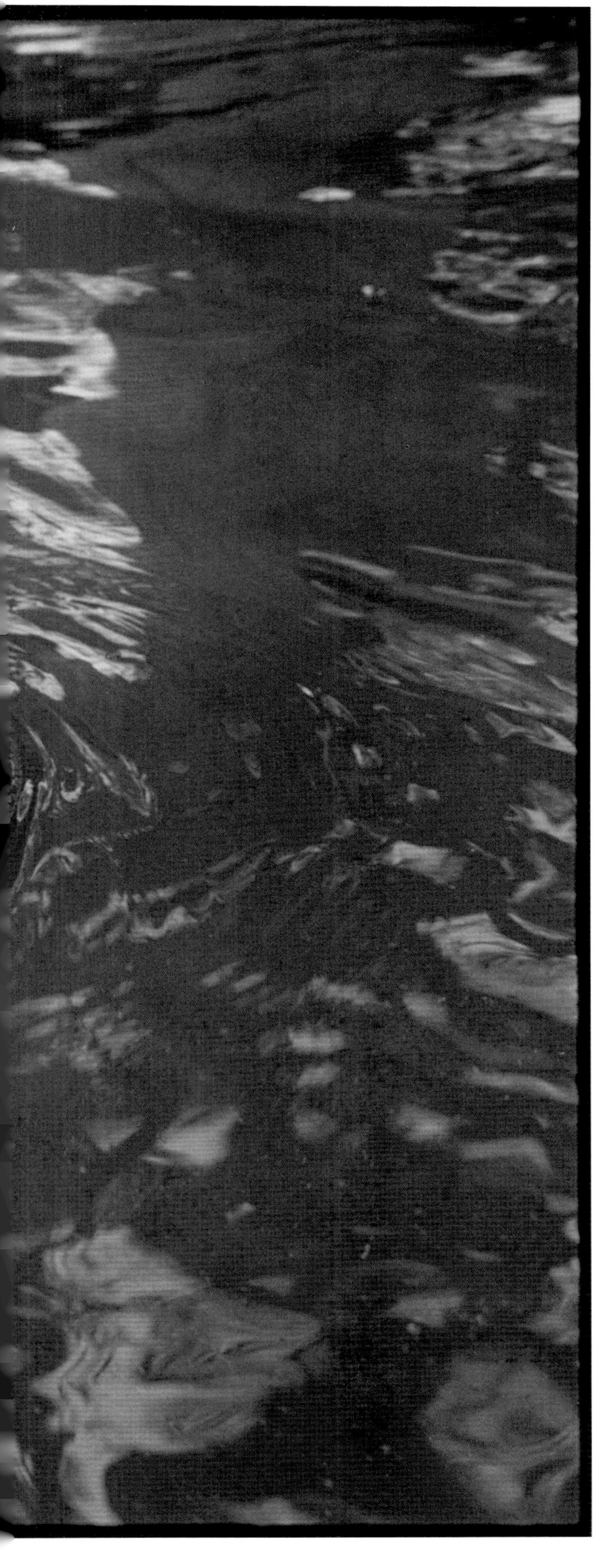

PLATE 7

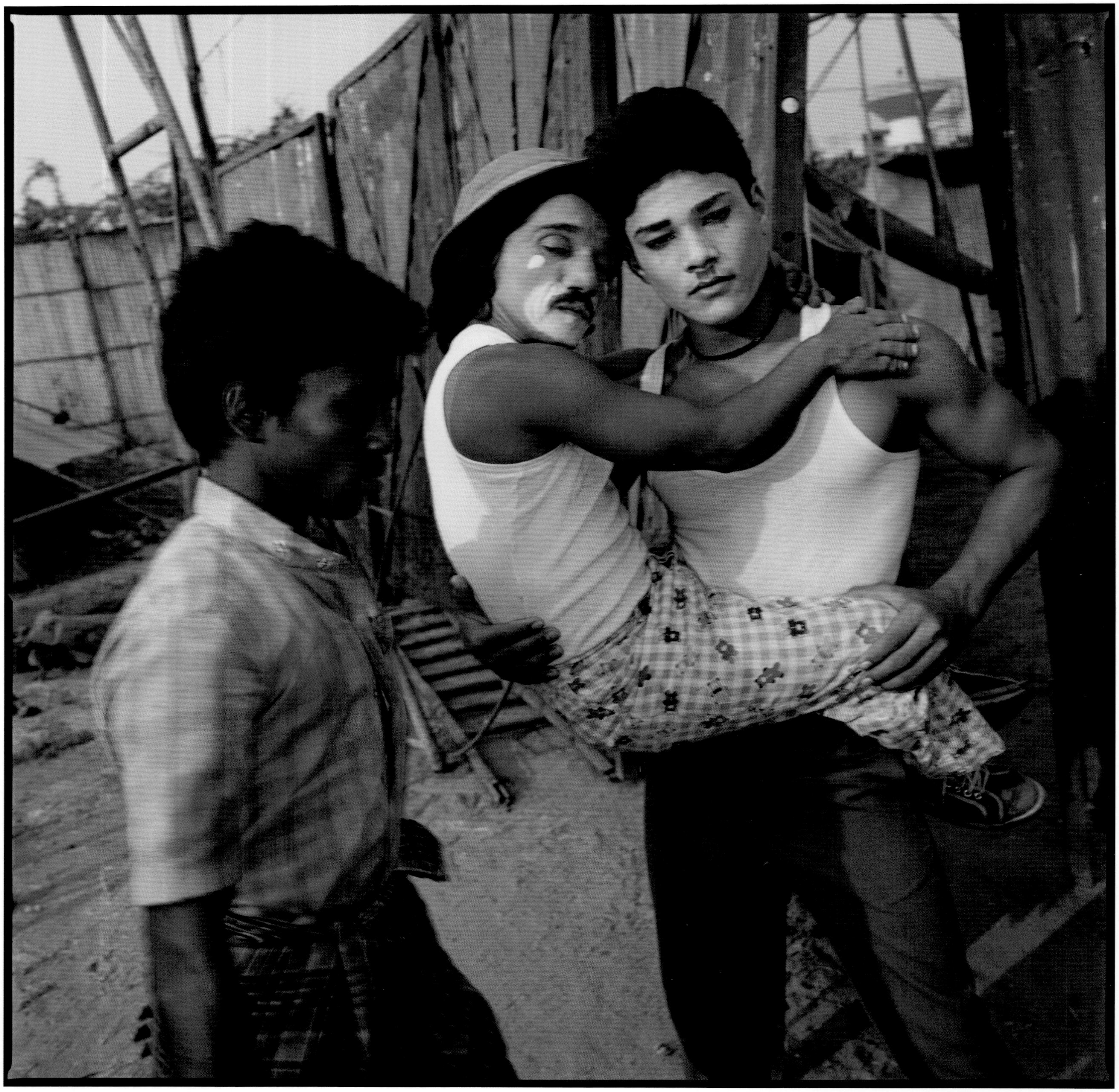

PLATE 8

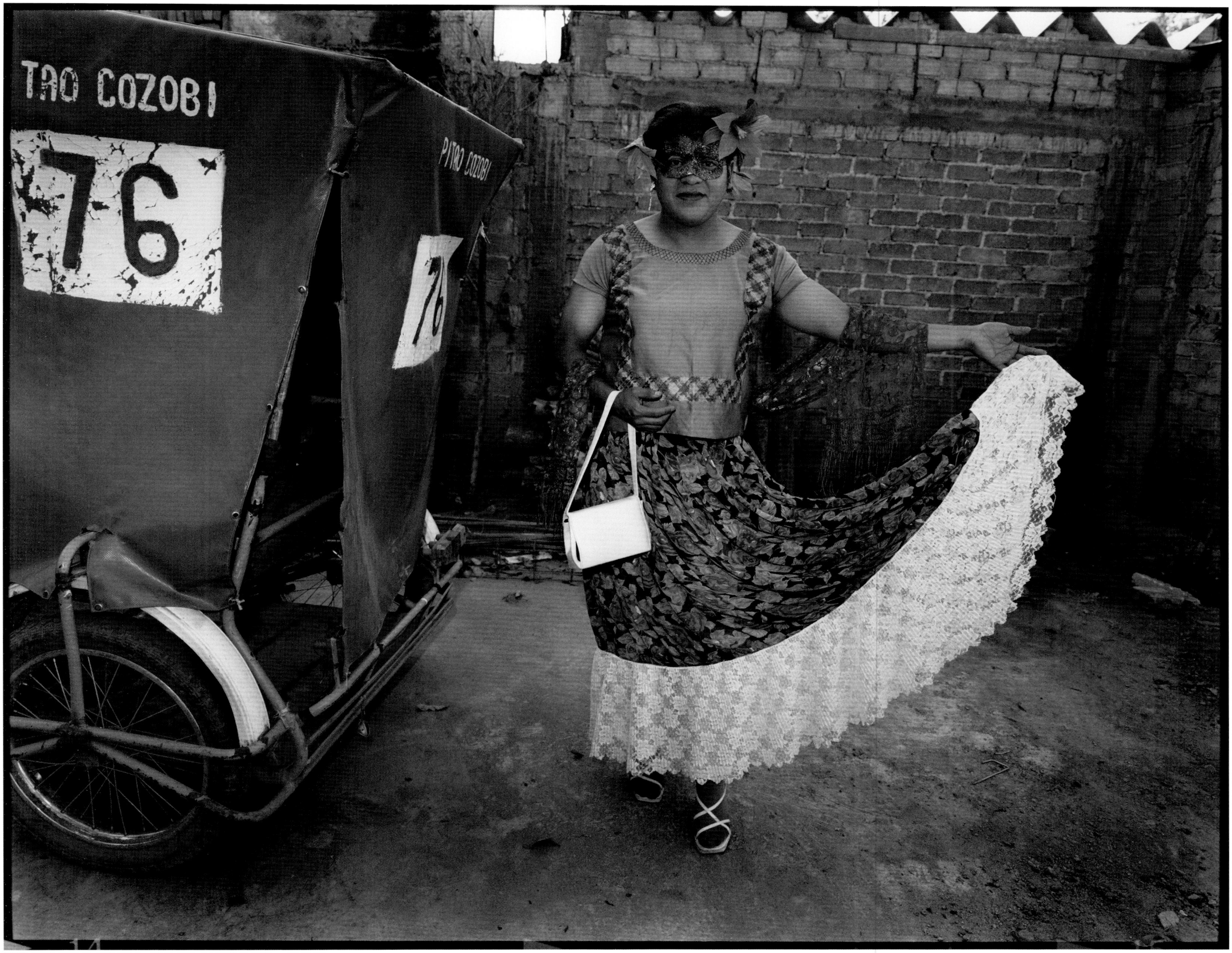

PLATE 9

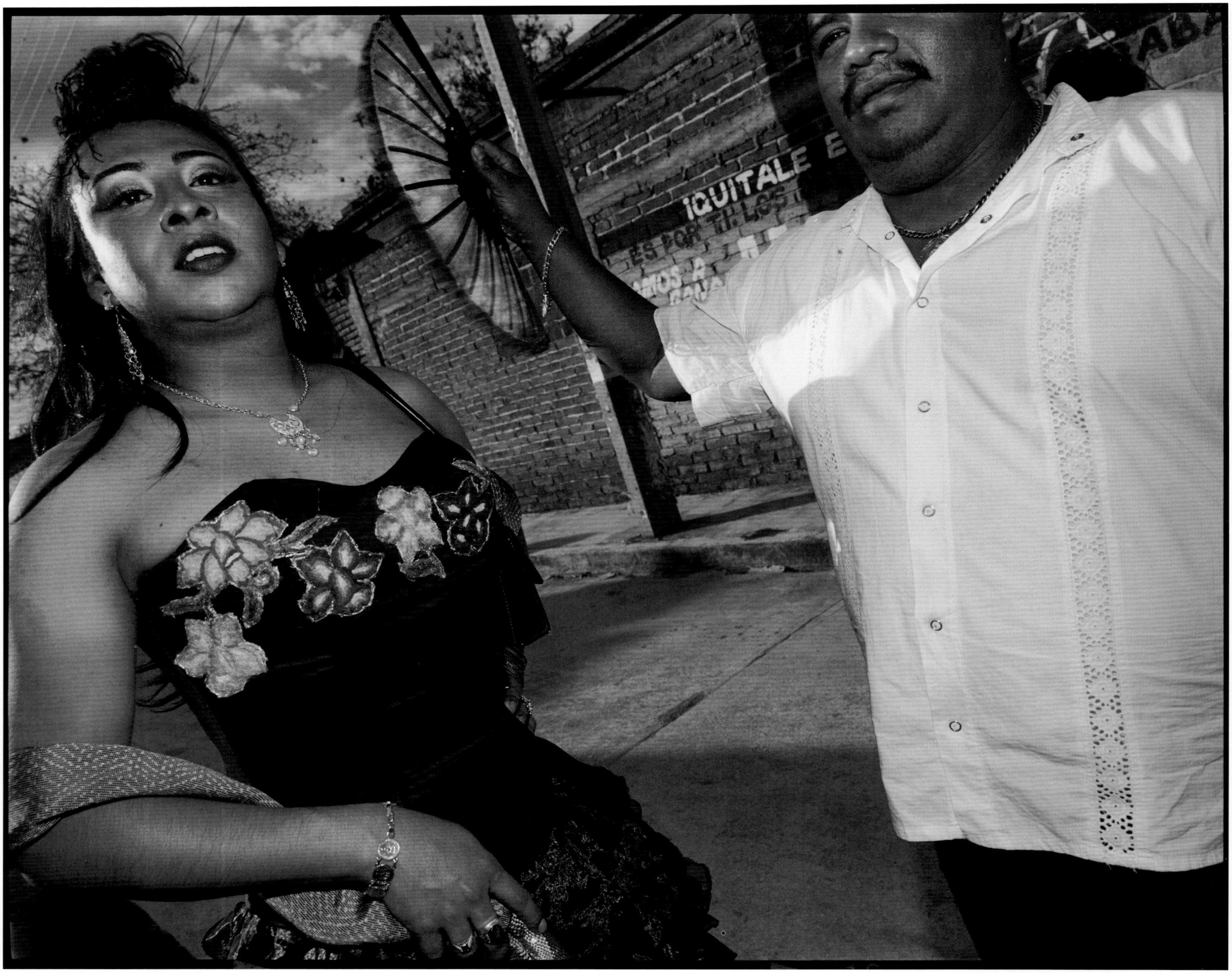

PLATE 10

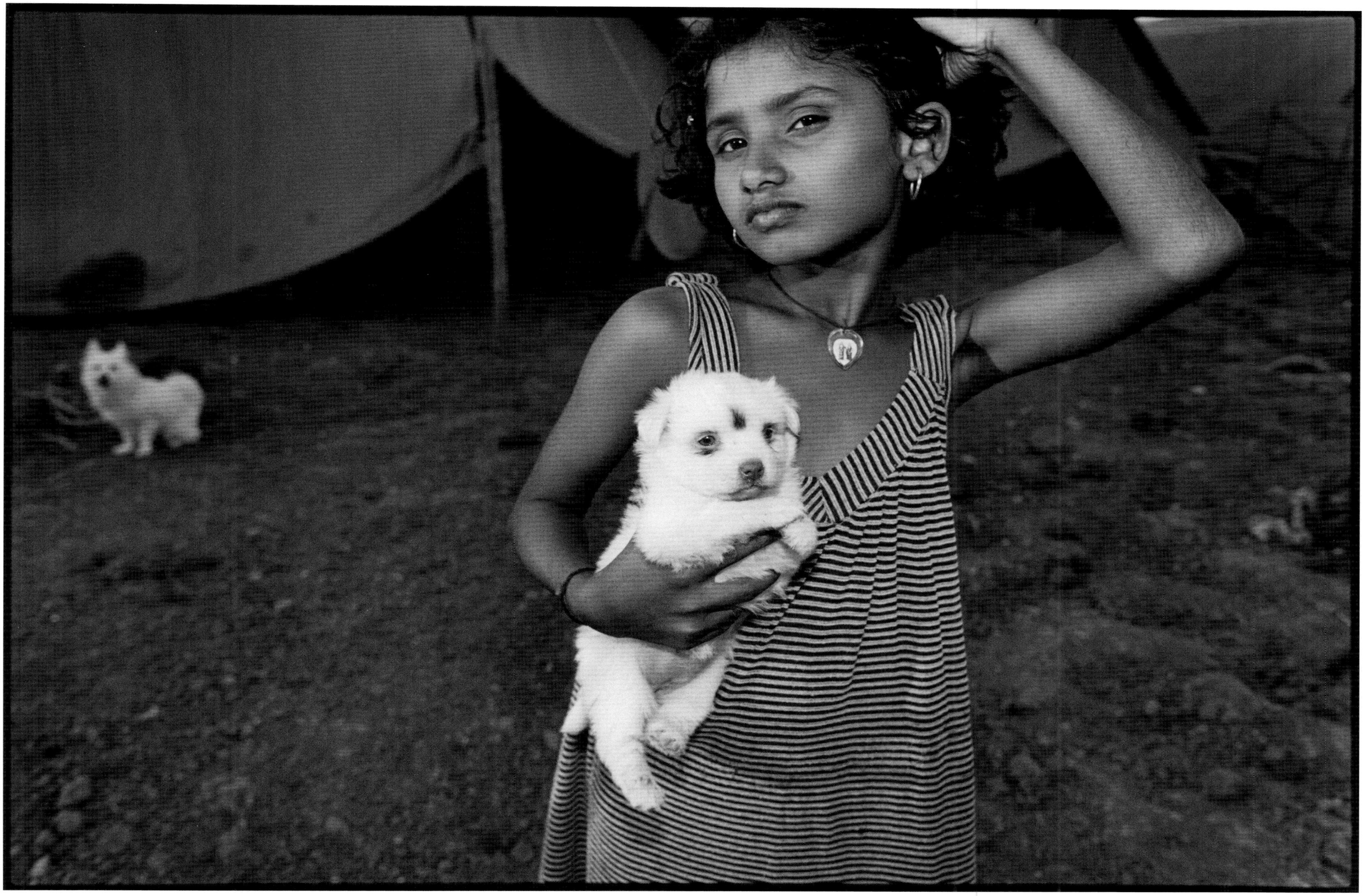

PLATE 11

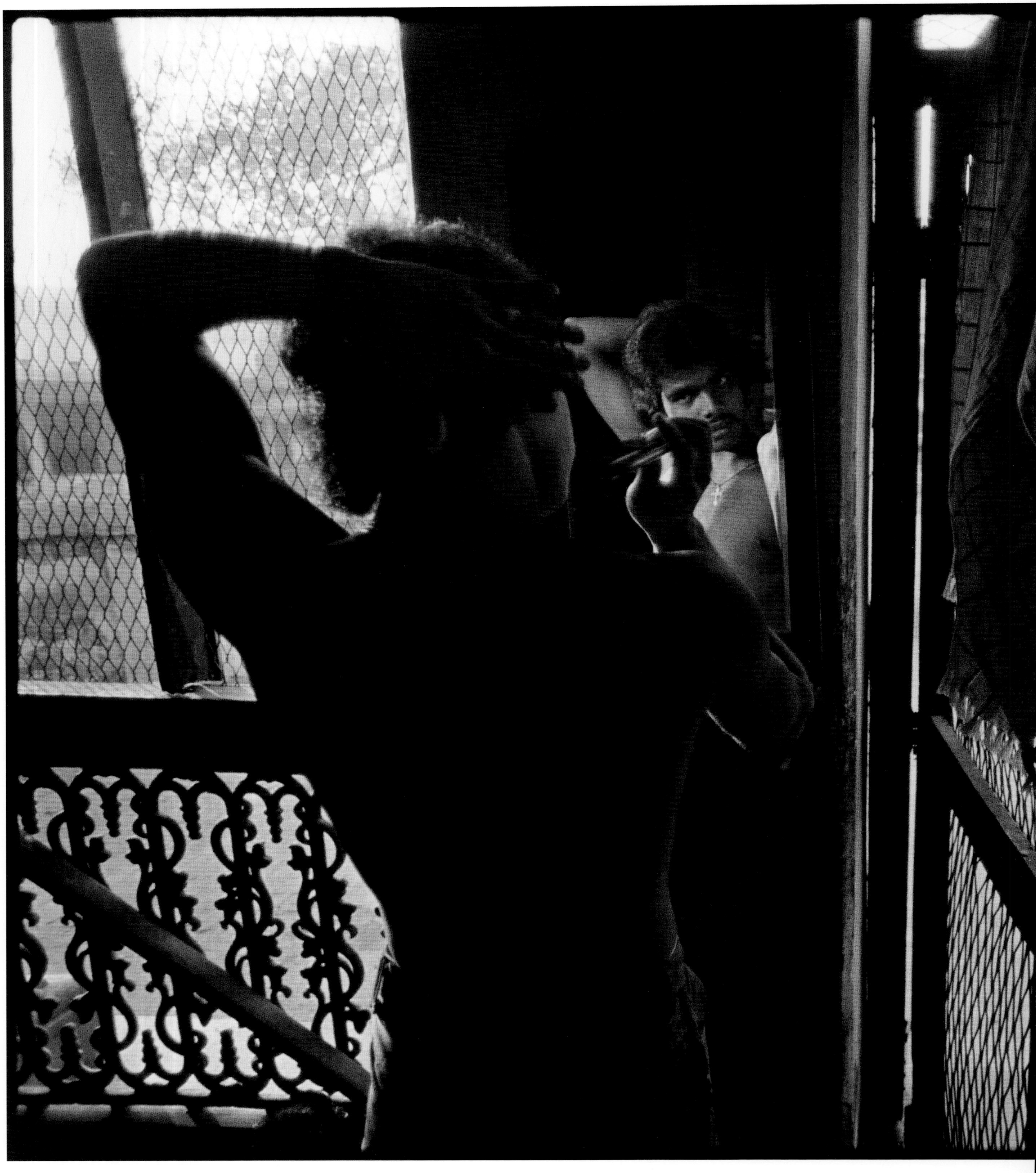

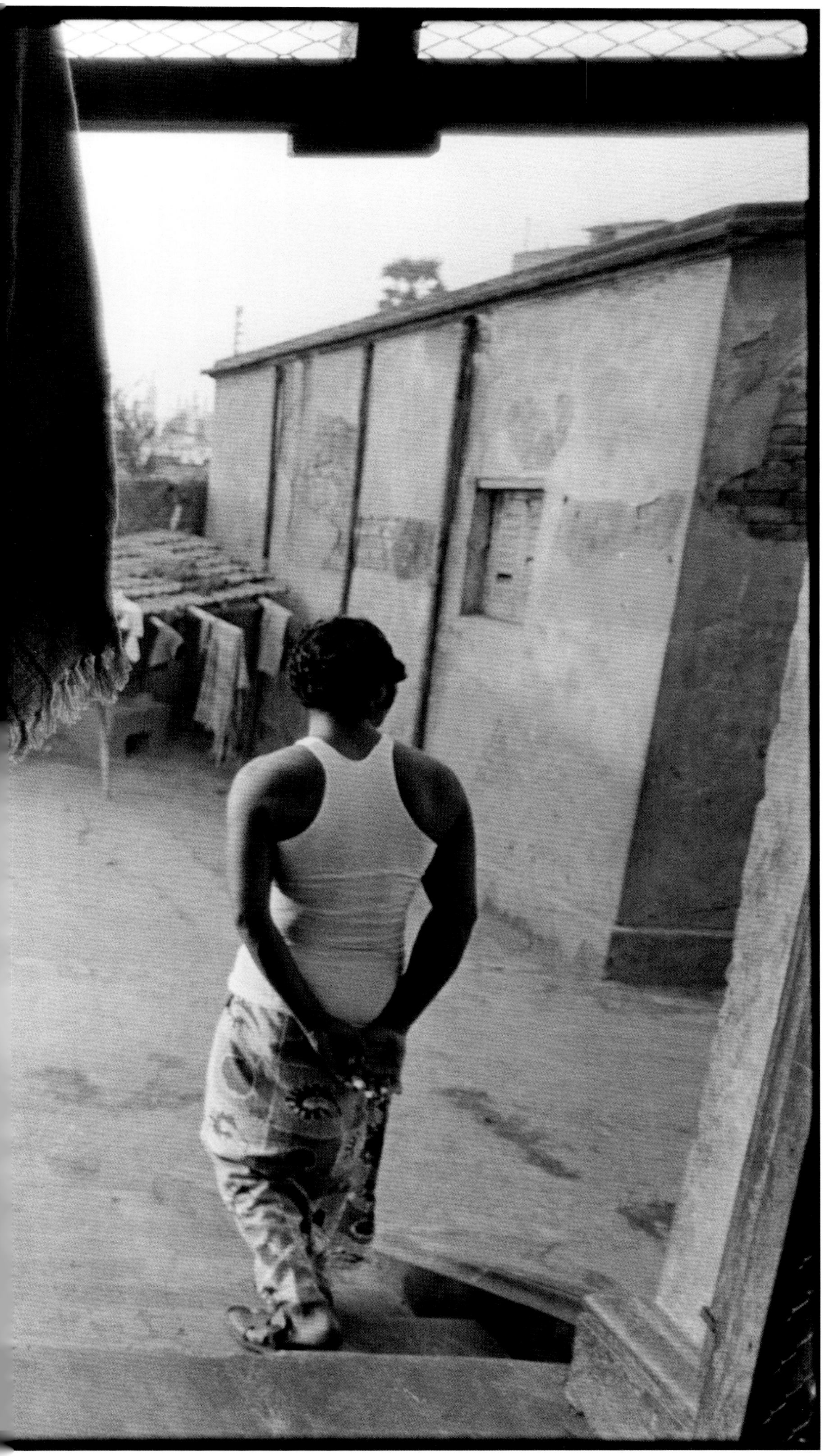

PLATE 12

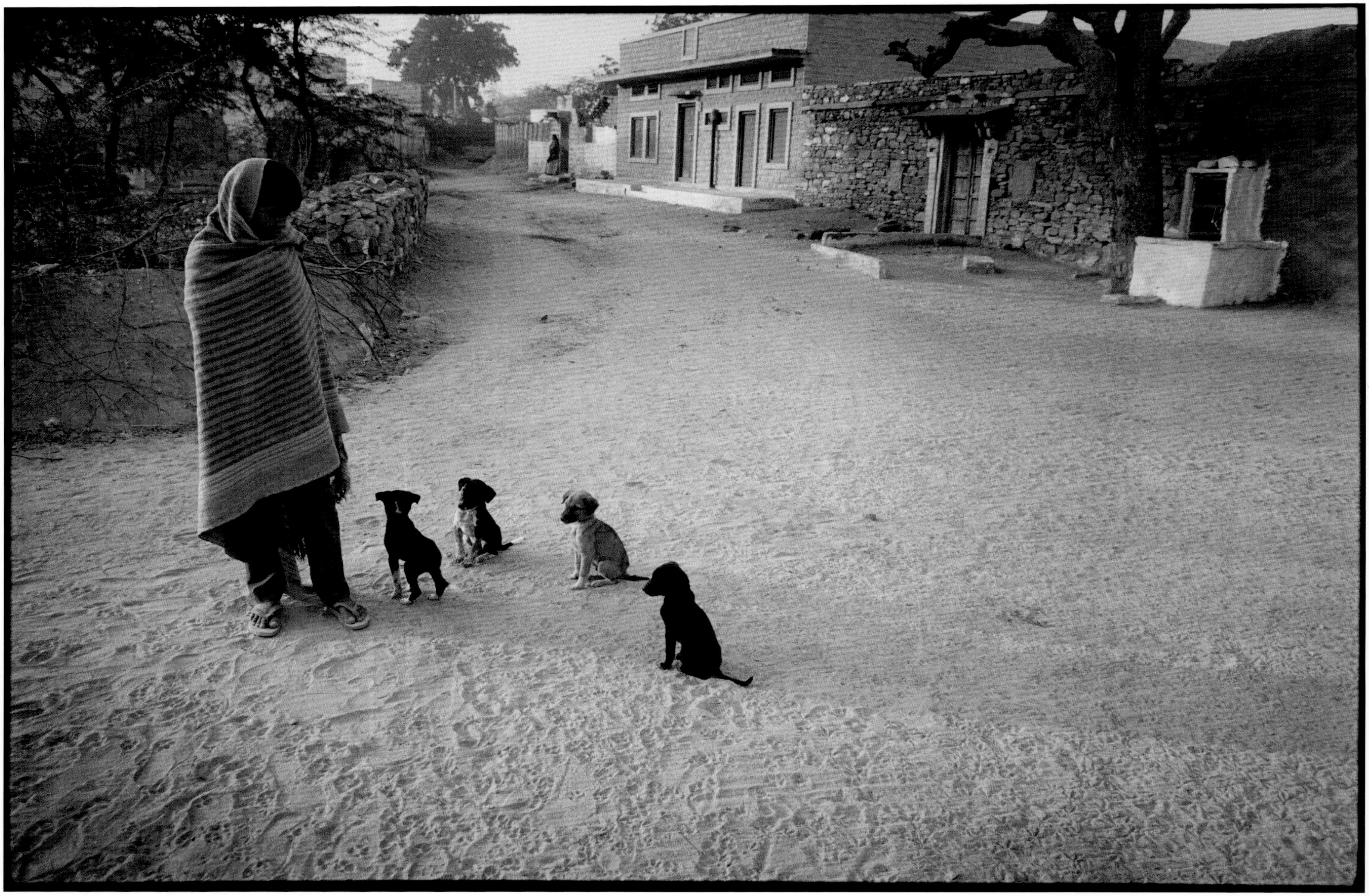

PLATE 13

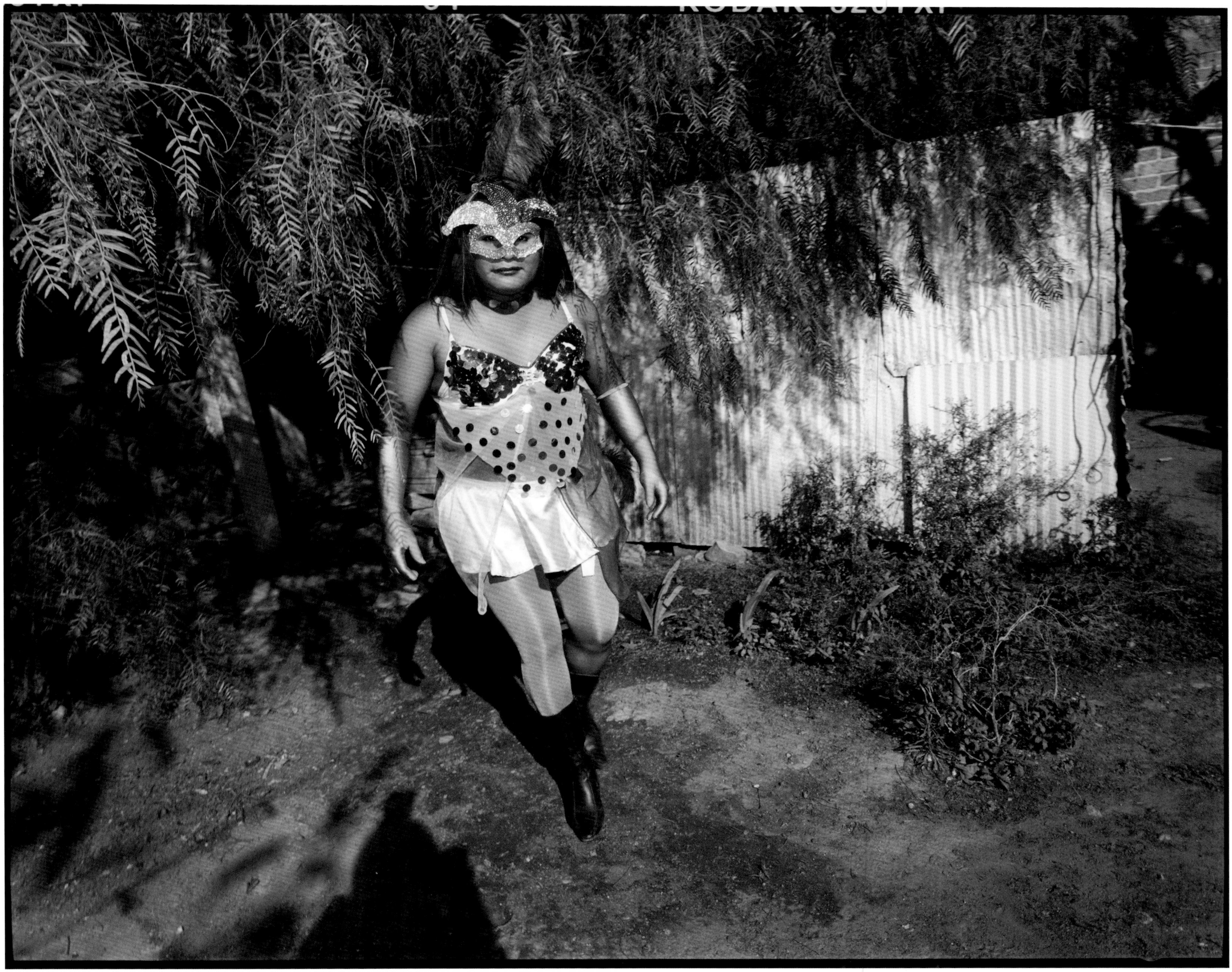

PLATE 14

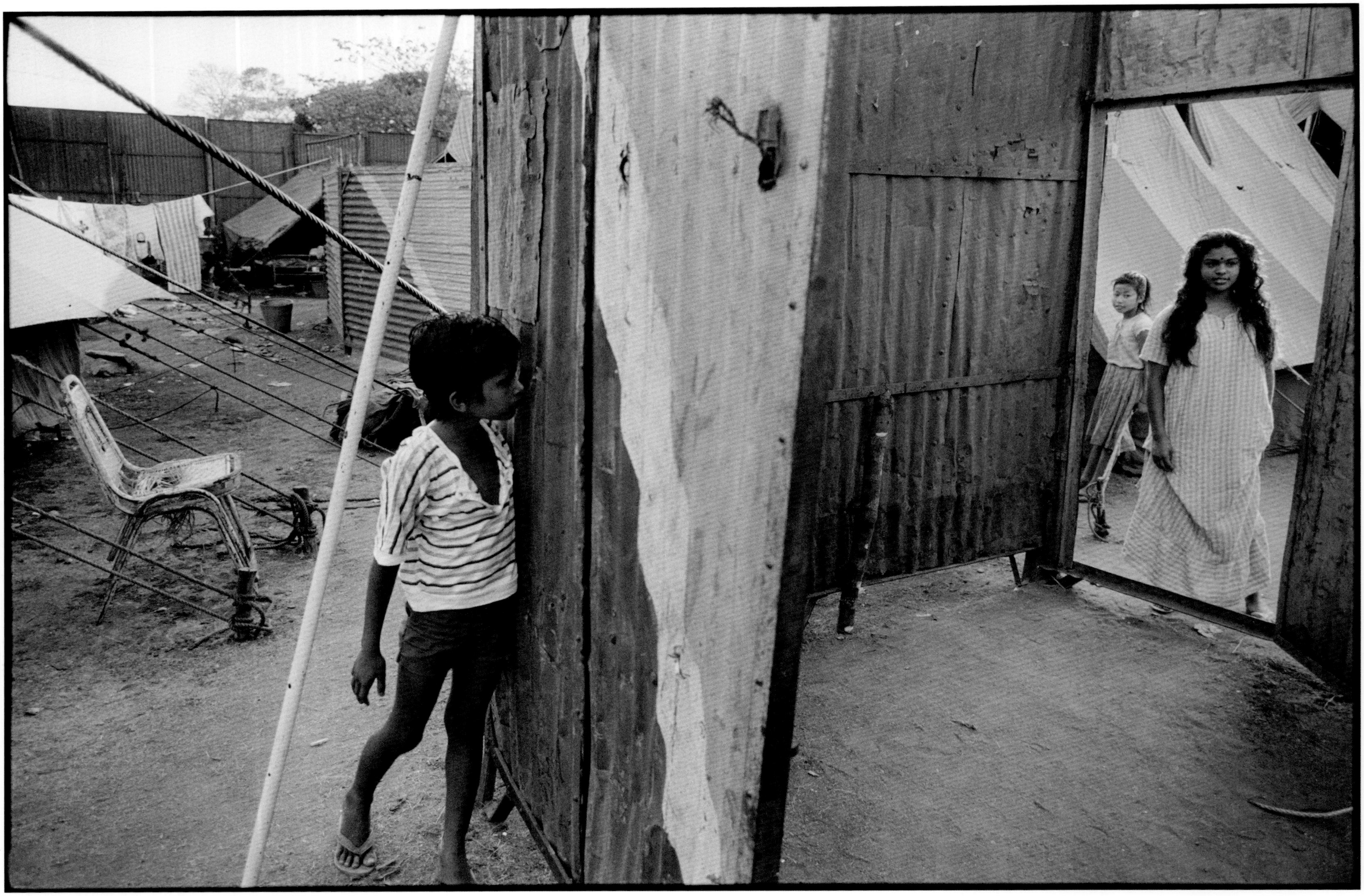

PLATE 15

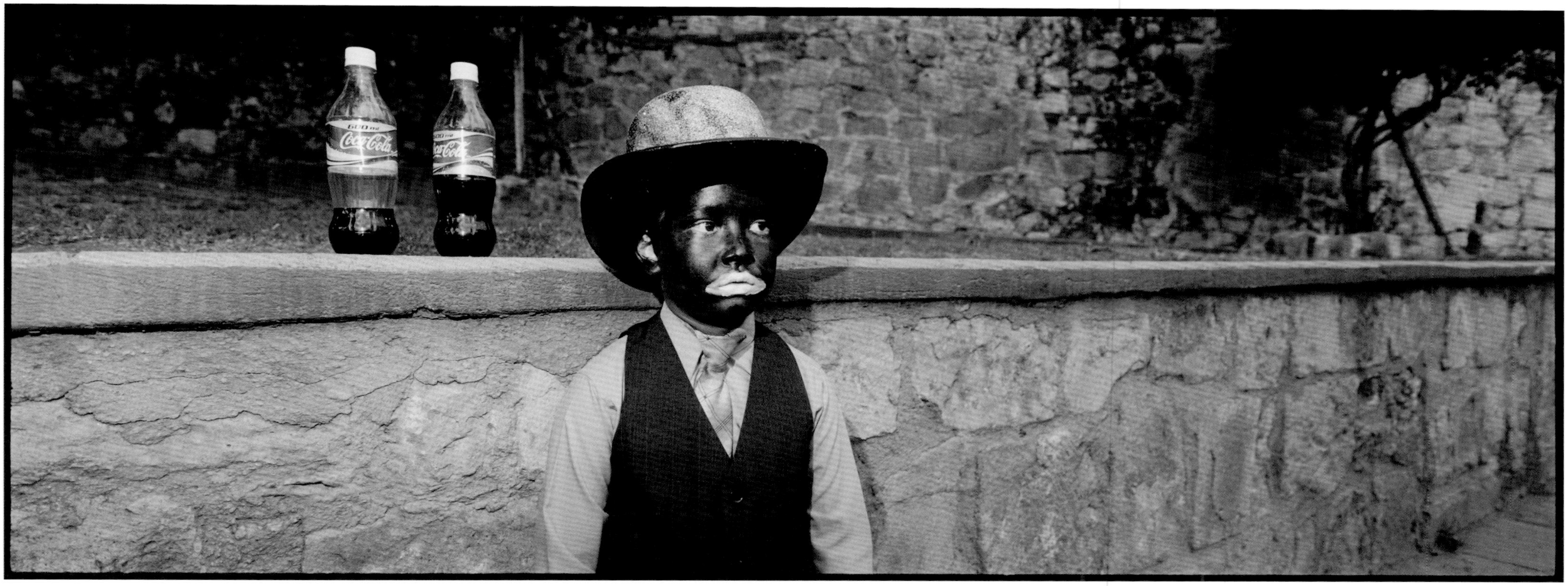

PLATE 16

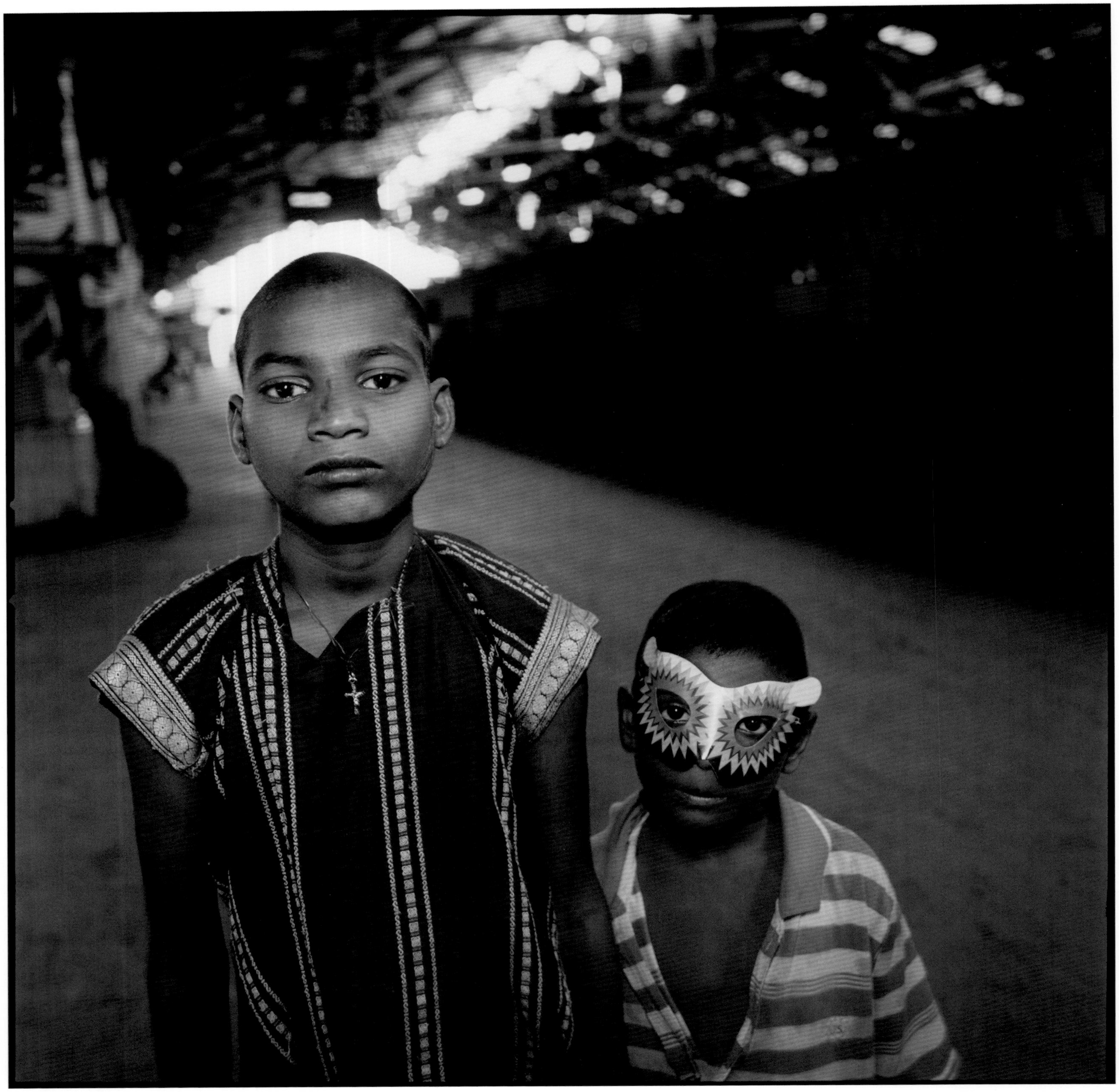

PLATE 17

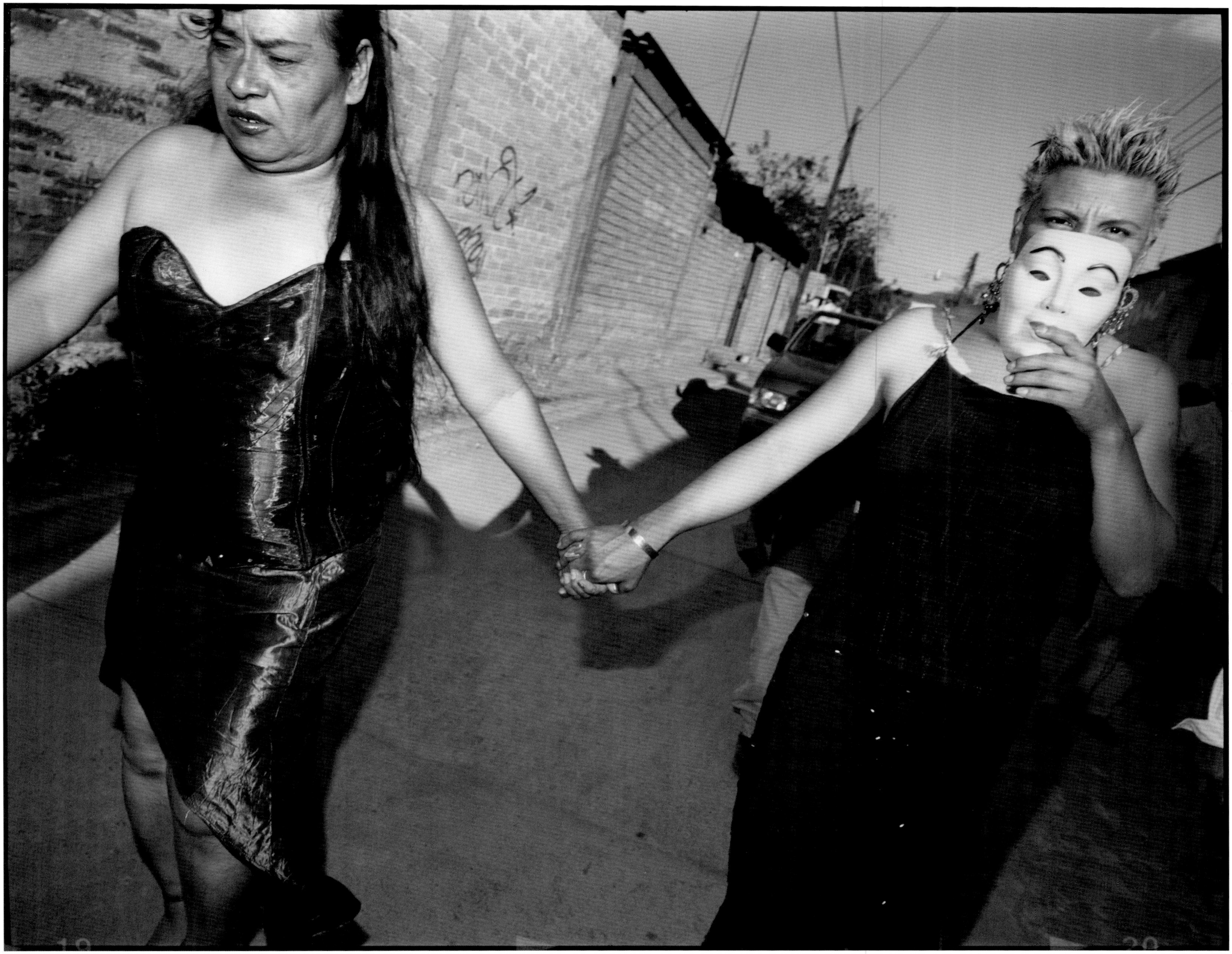

PLATE 18

PLATE 19

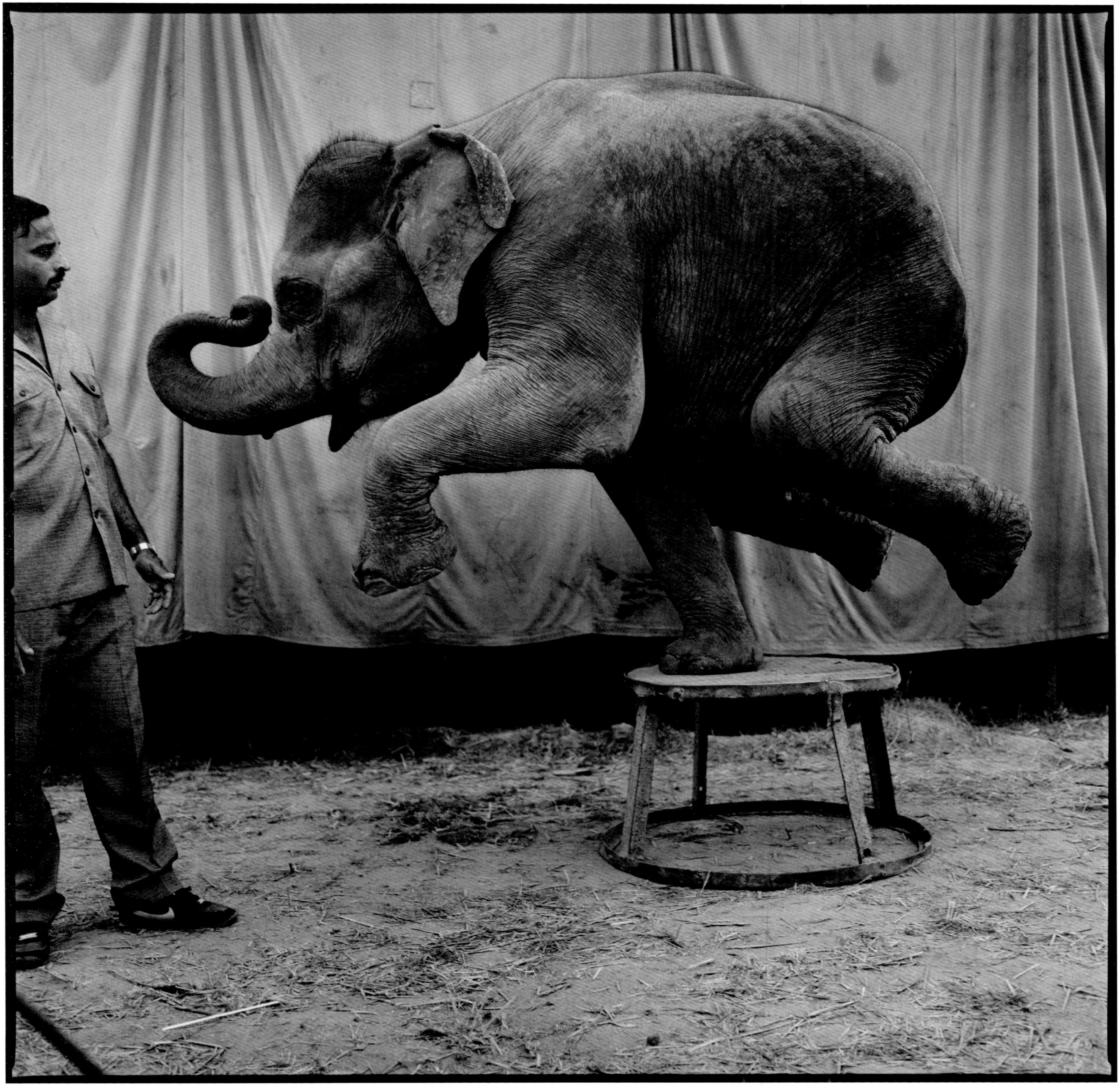

PLATE 20

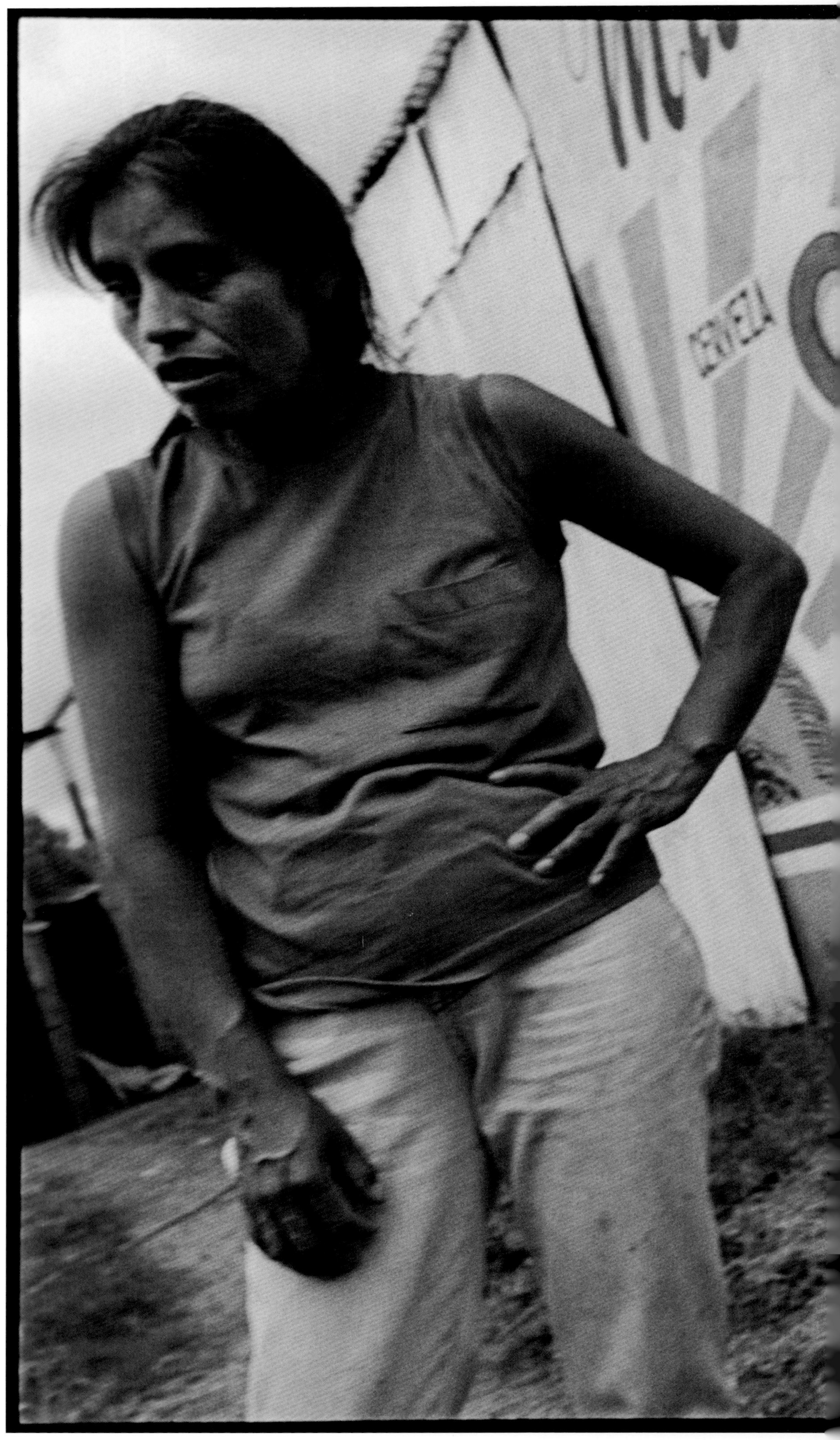

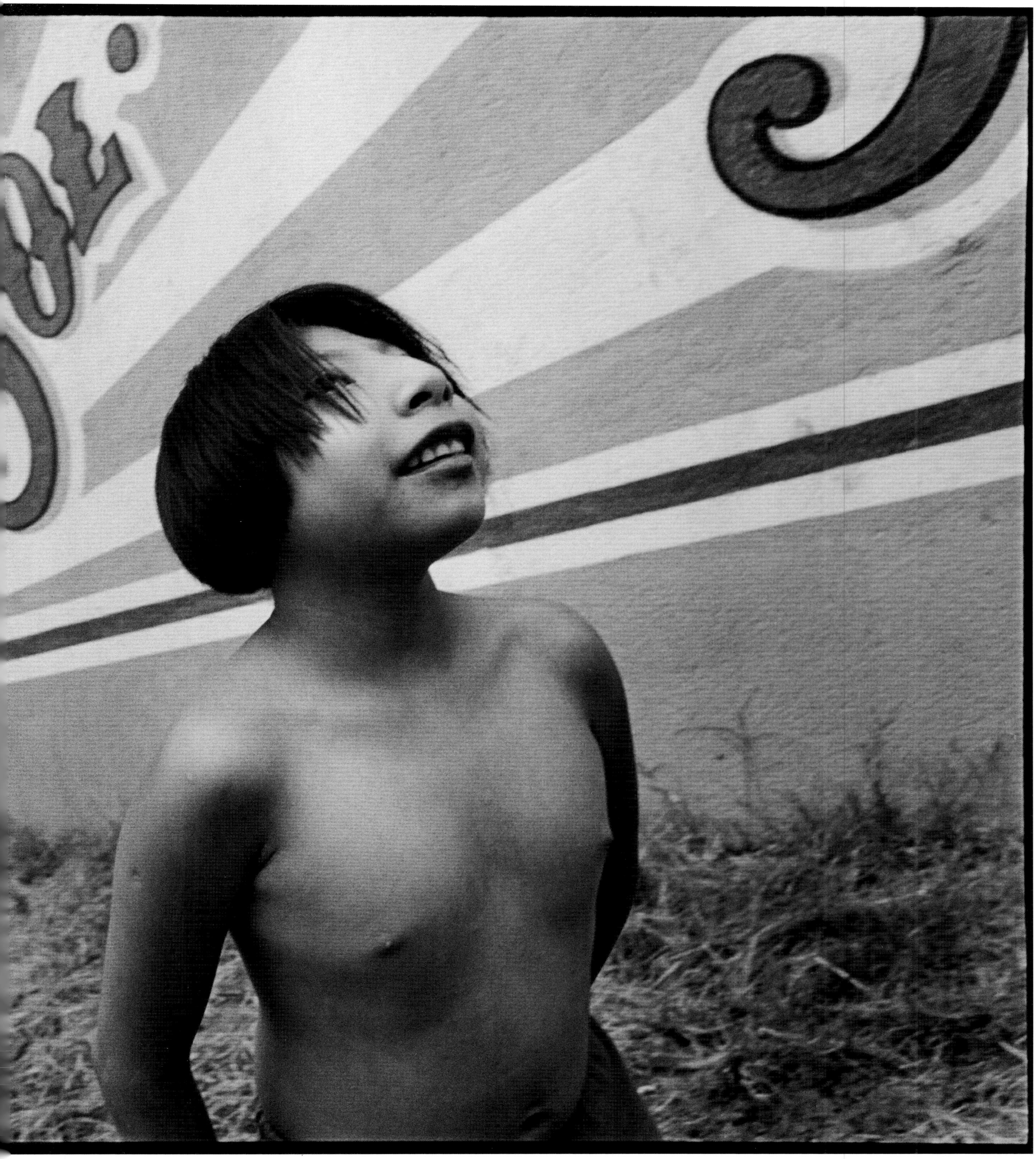

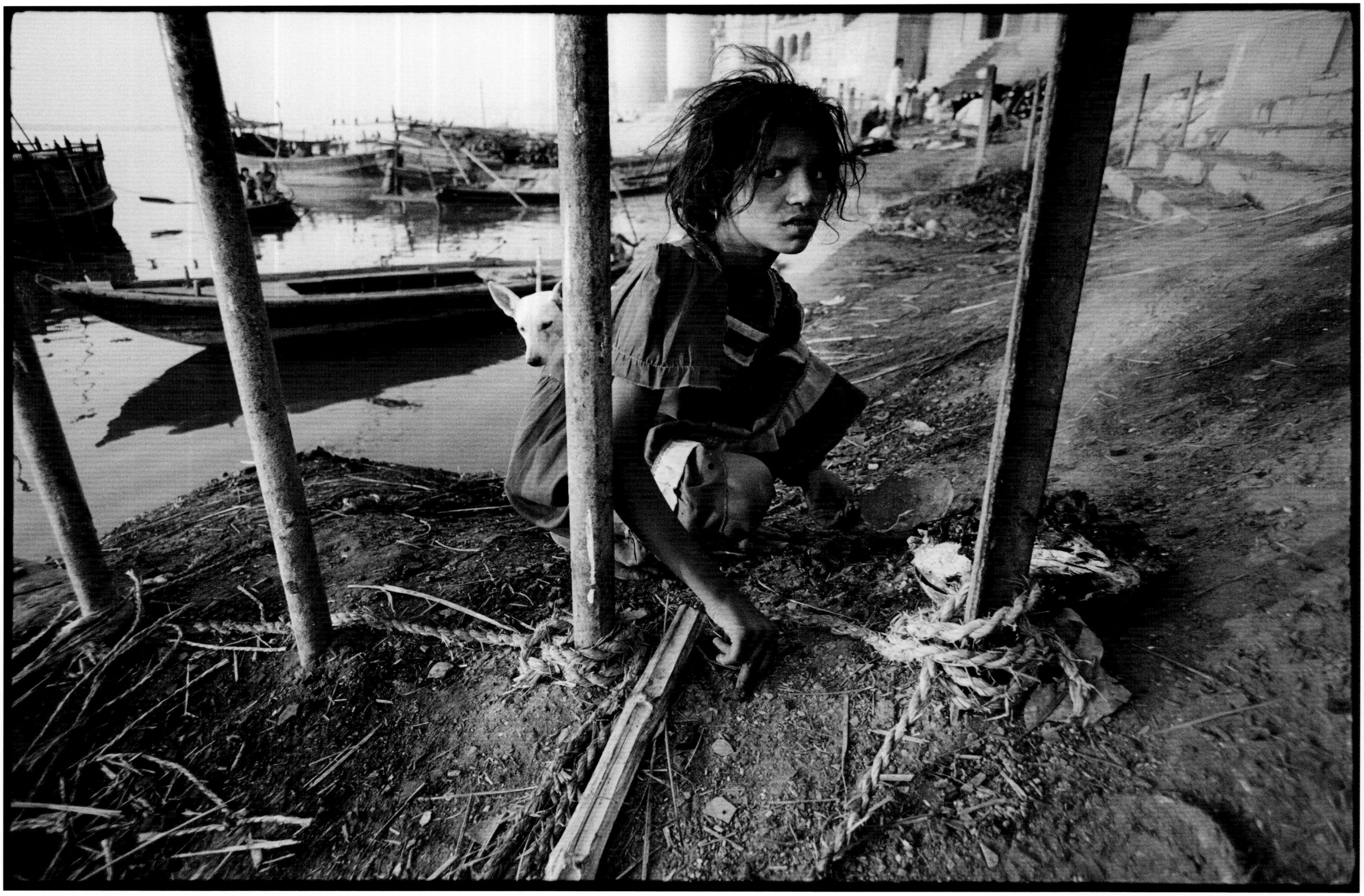

PLATE 21

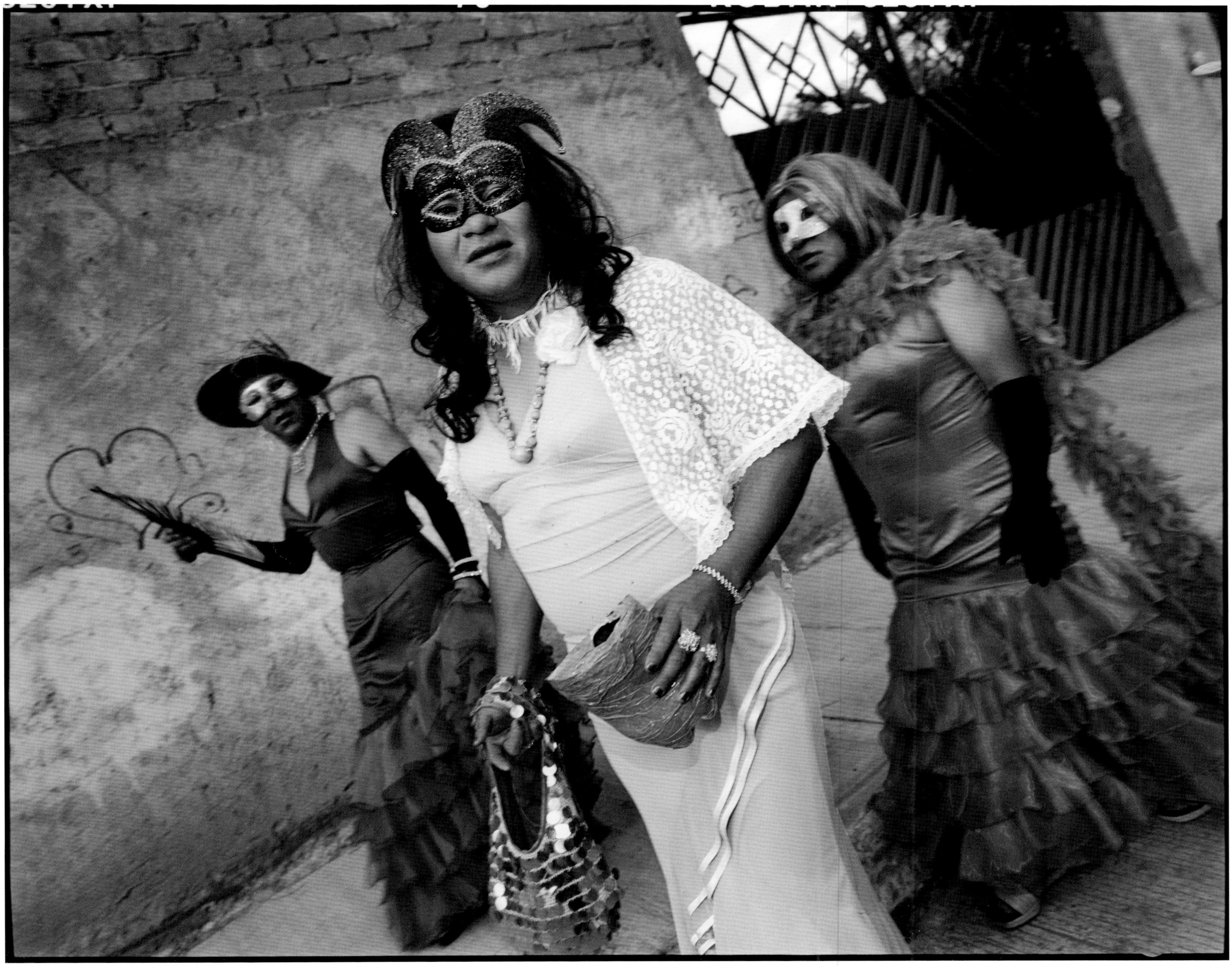

PLATE 22

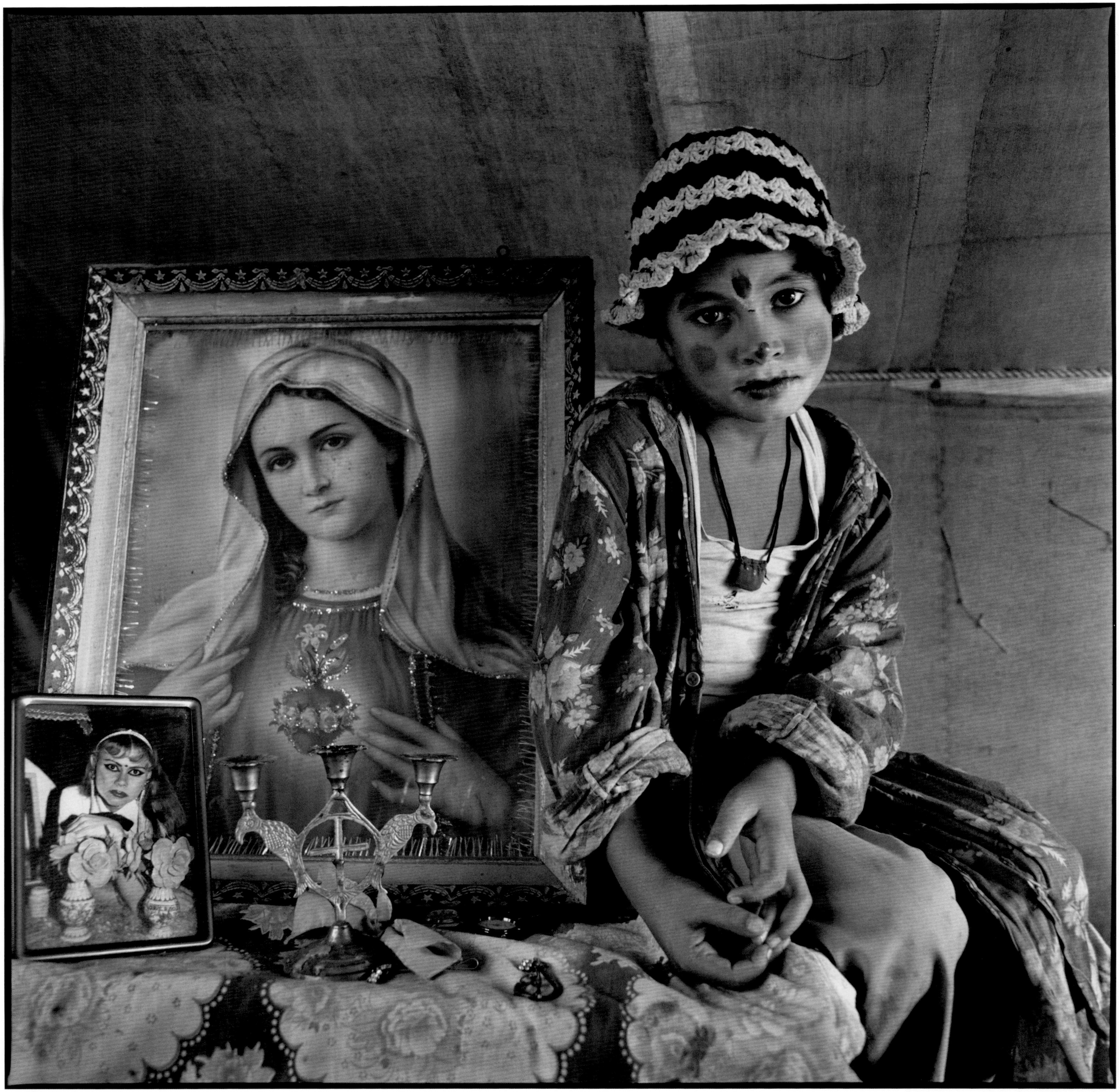

PLATE 23

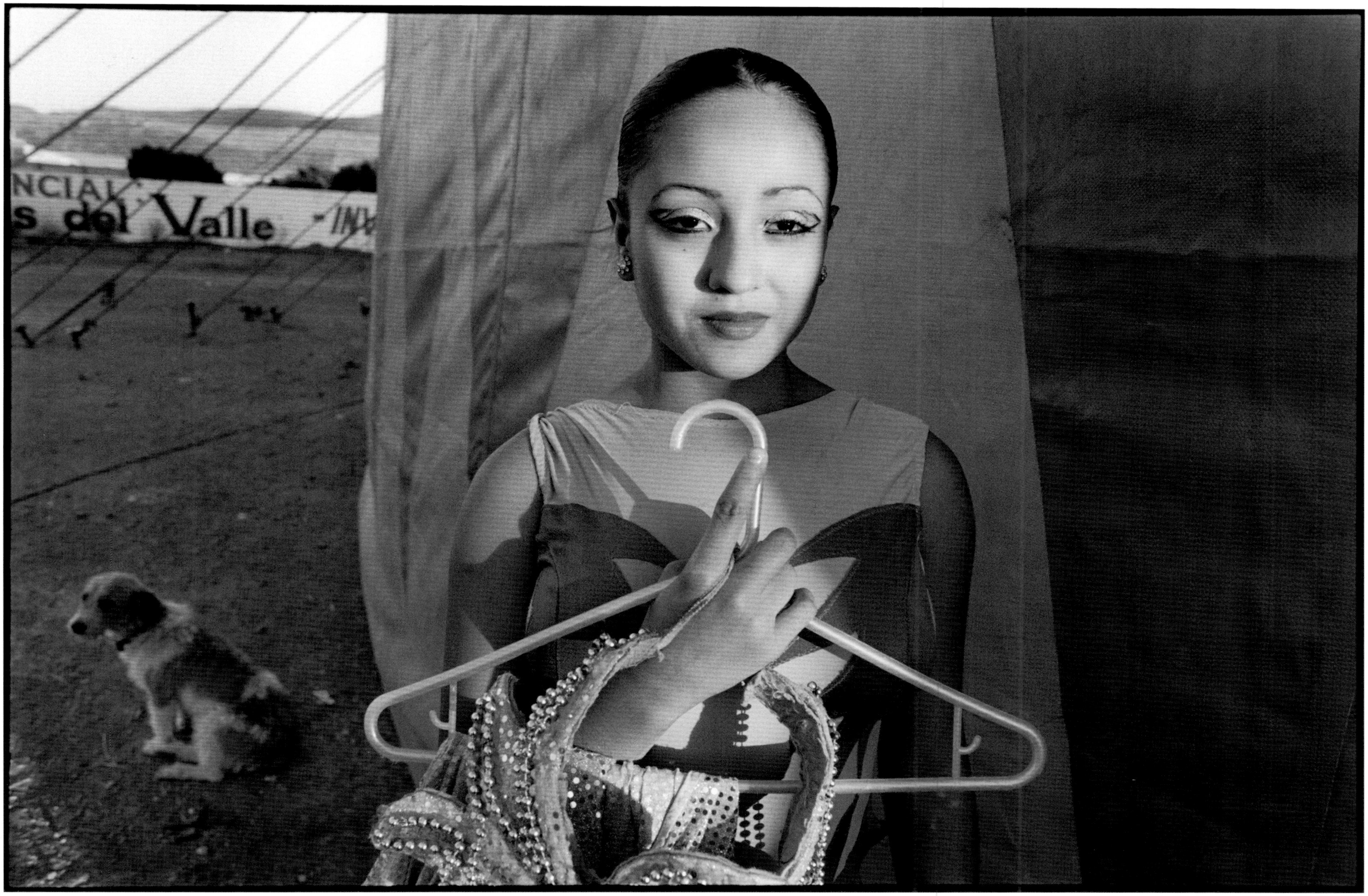

PLATE 24

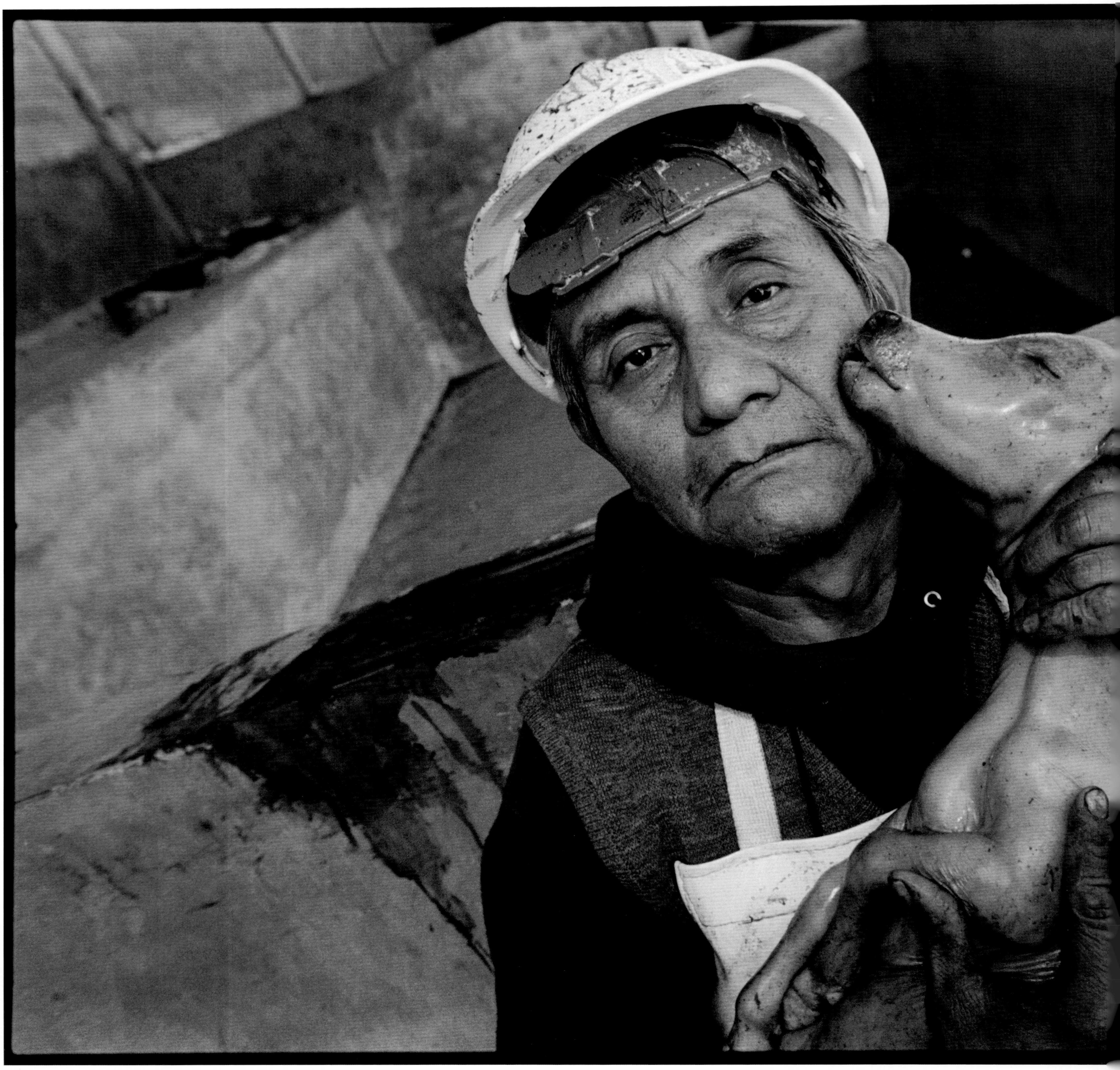

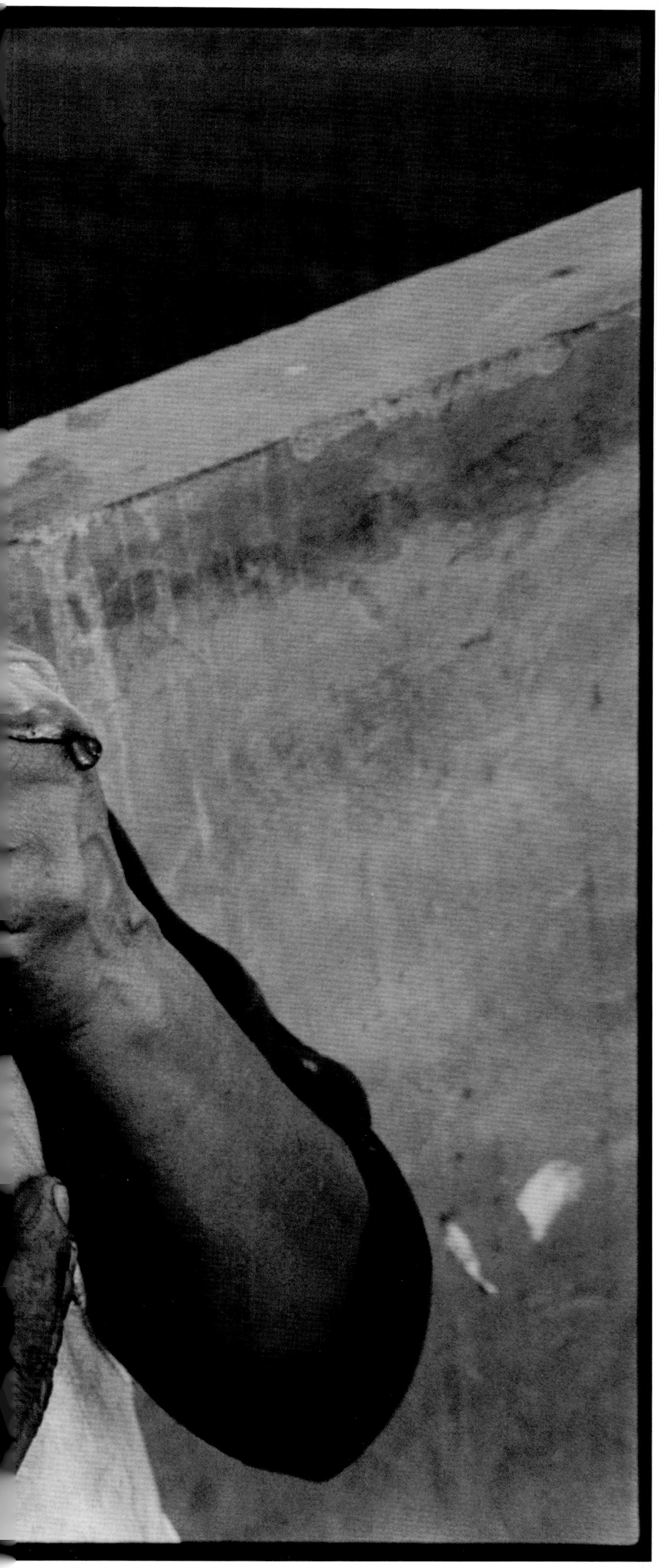

PLATE 25

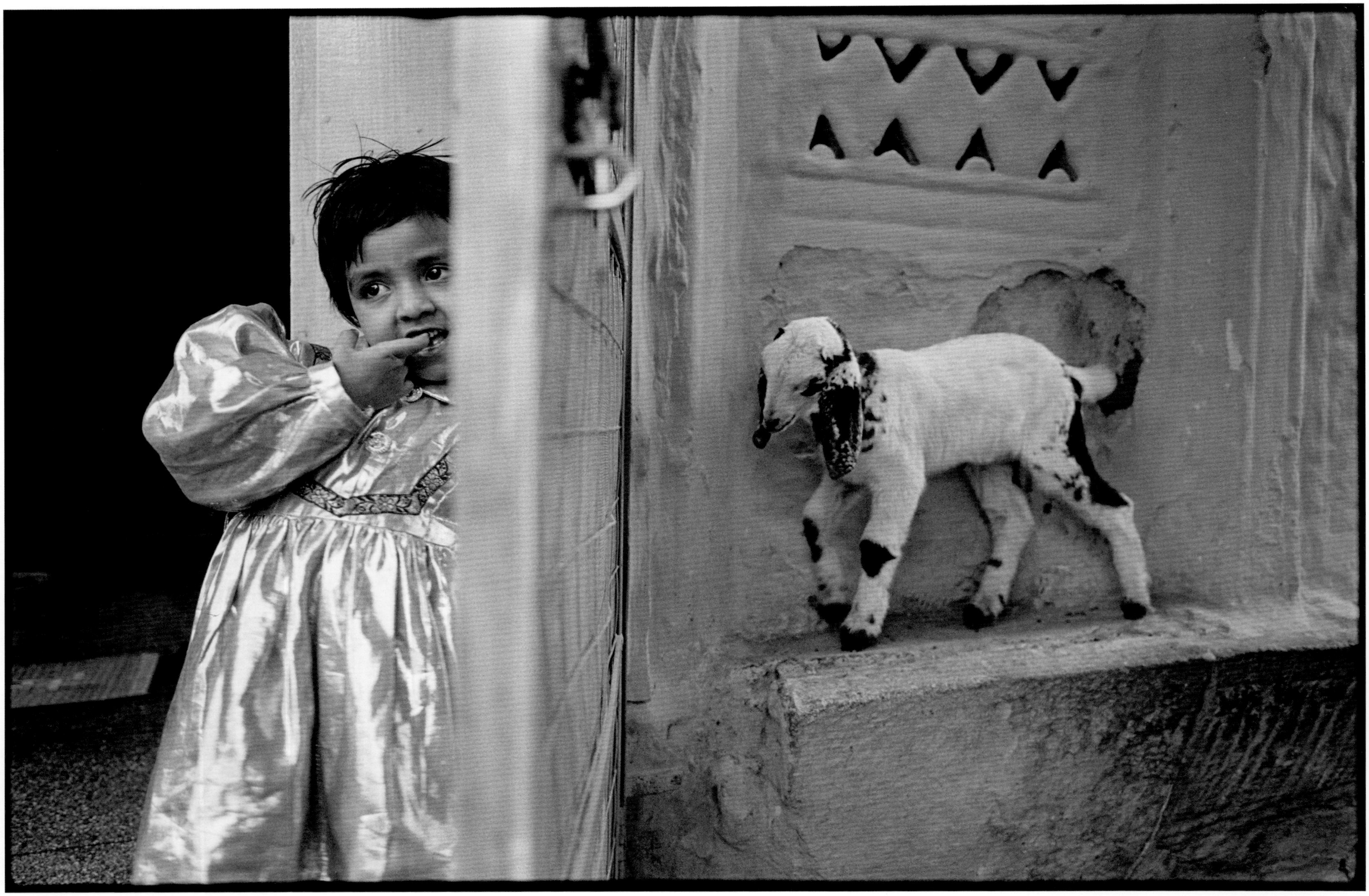

PLATE 26

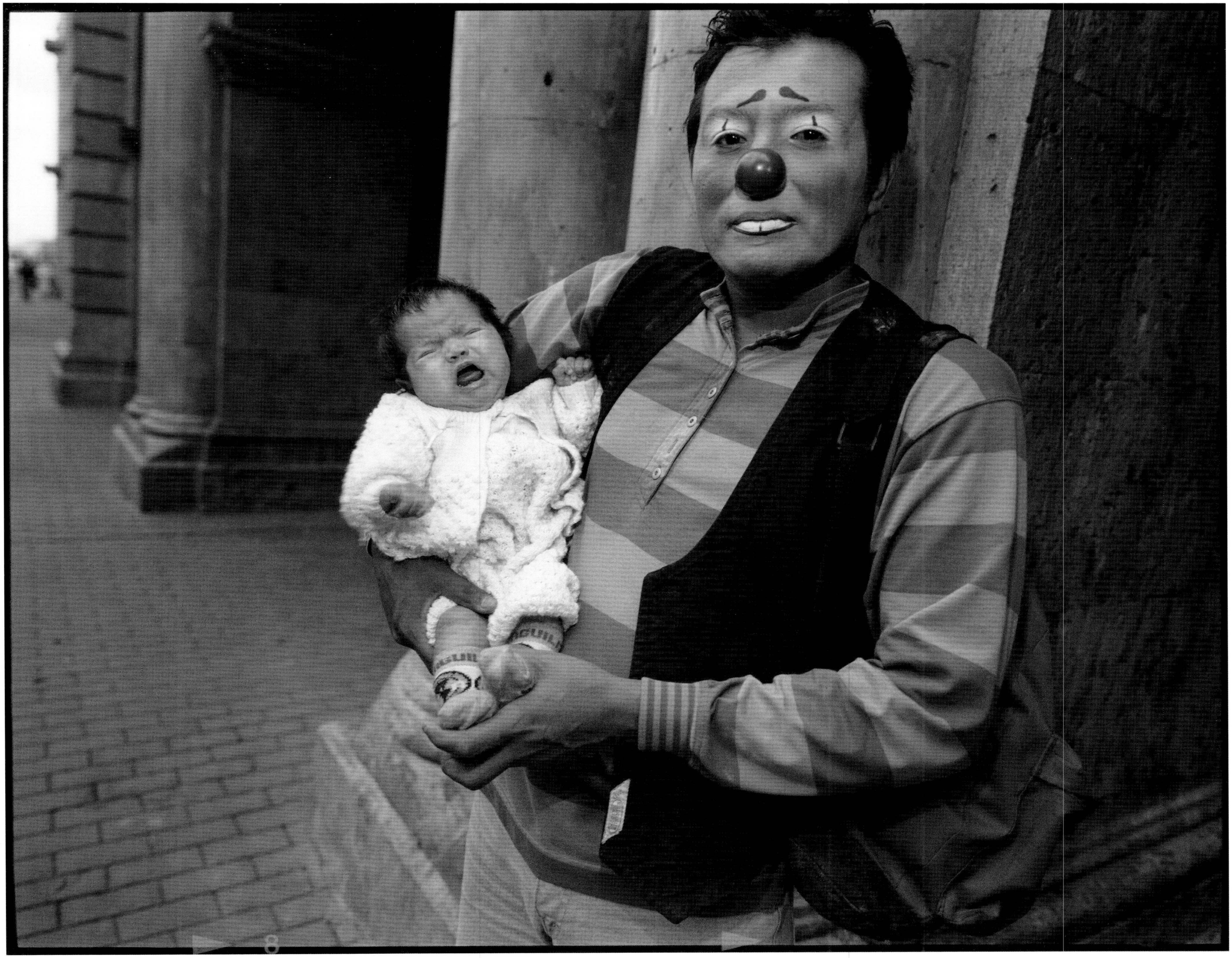

PLATE 27

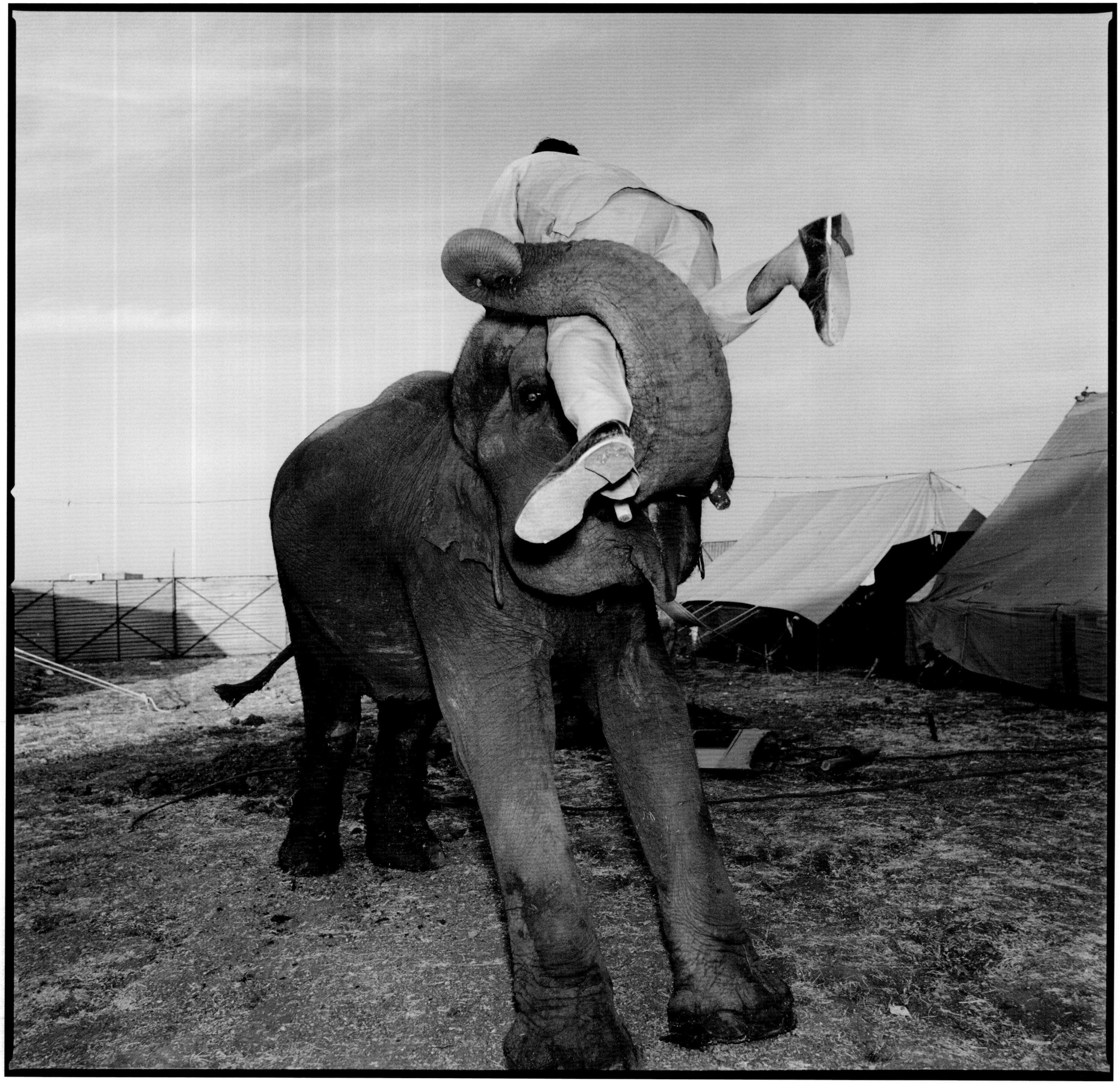

PLATE 28

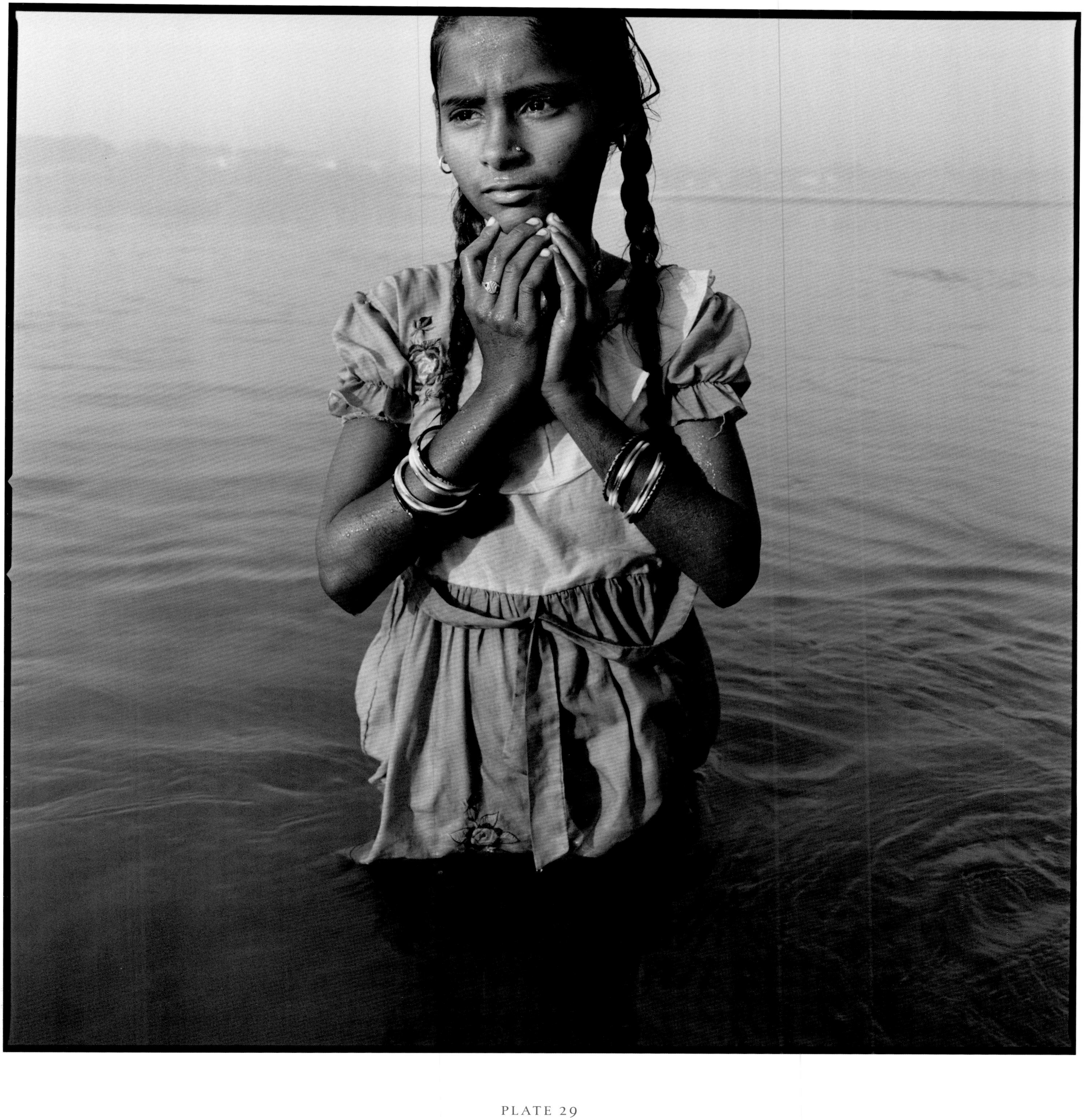

PLATE 29

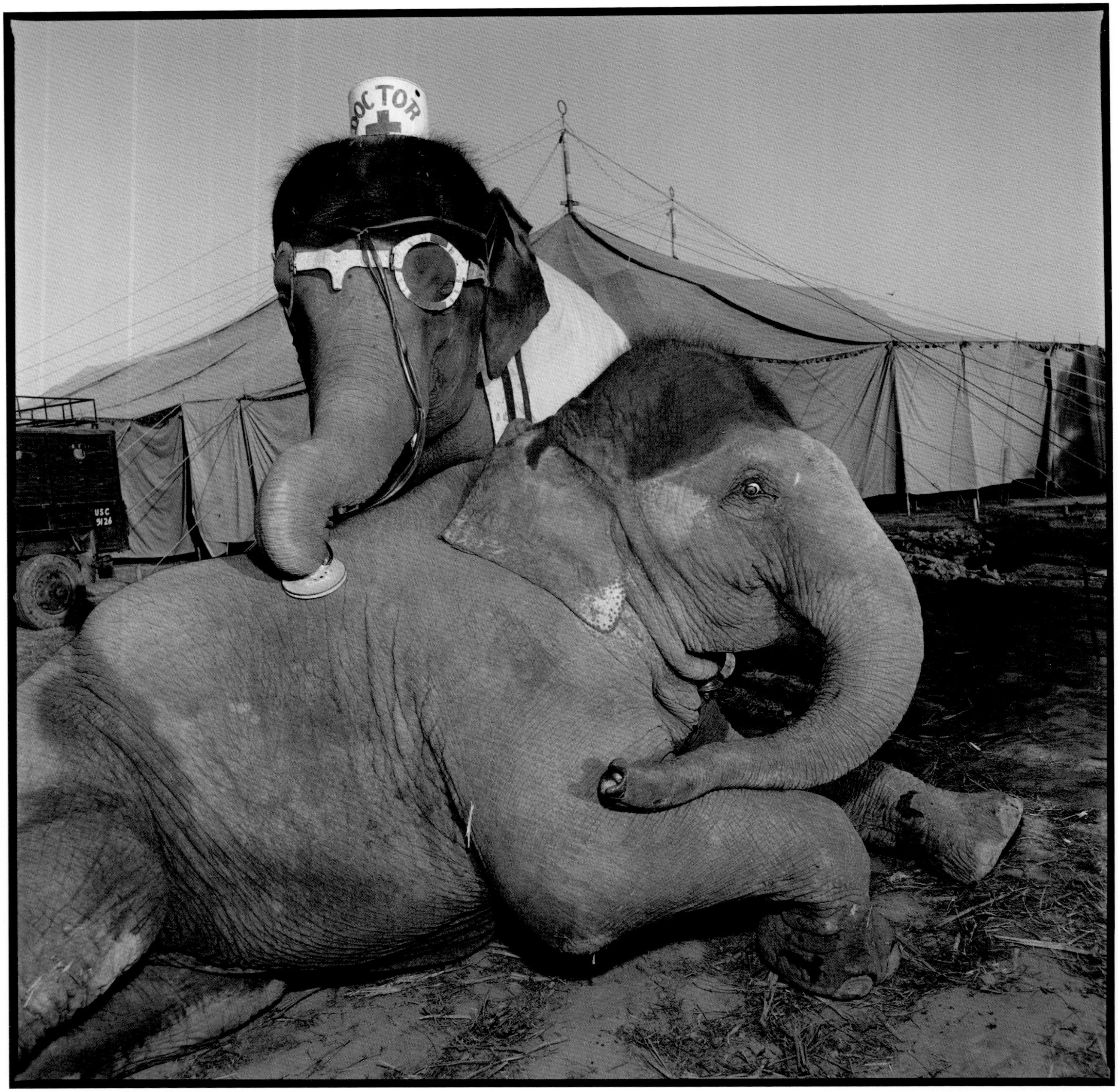

PLATE 30

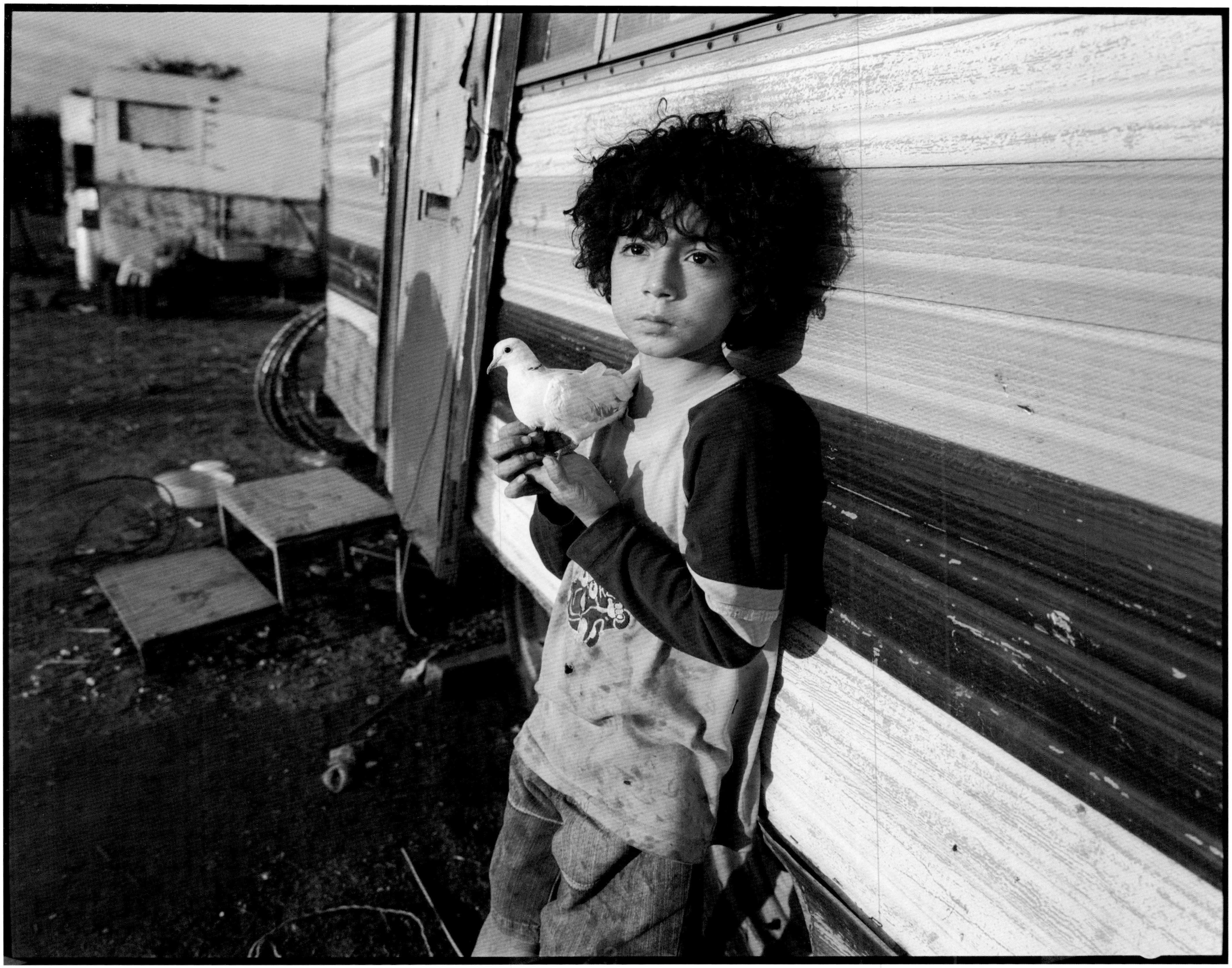

PLATE 31

PLATE 32

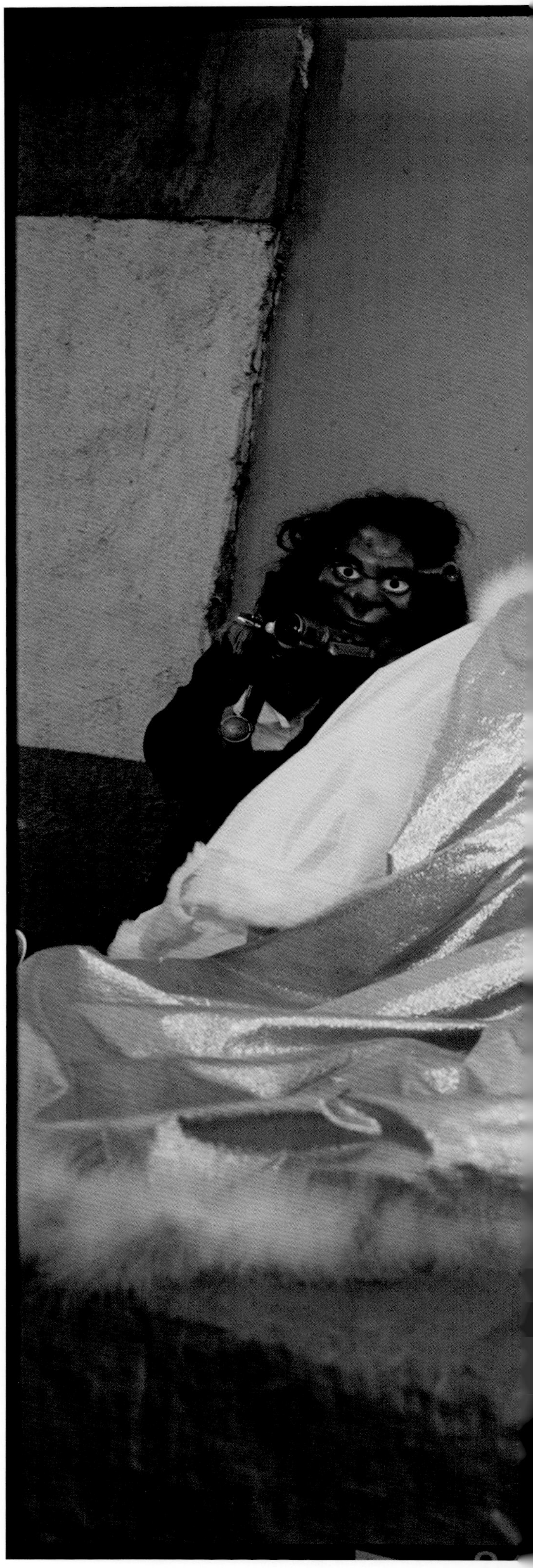

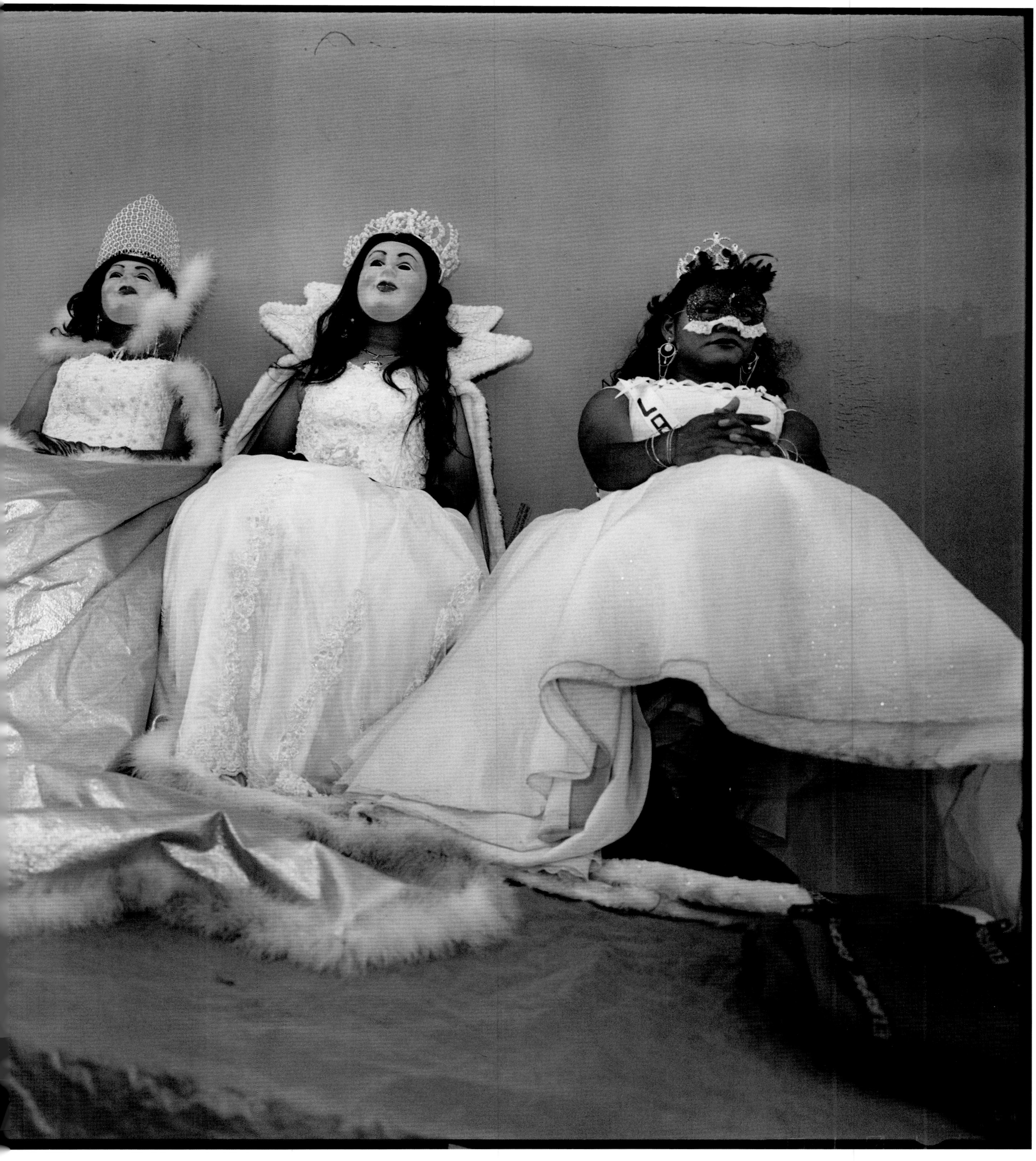

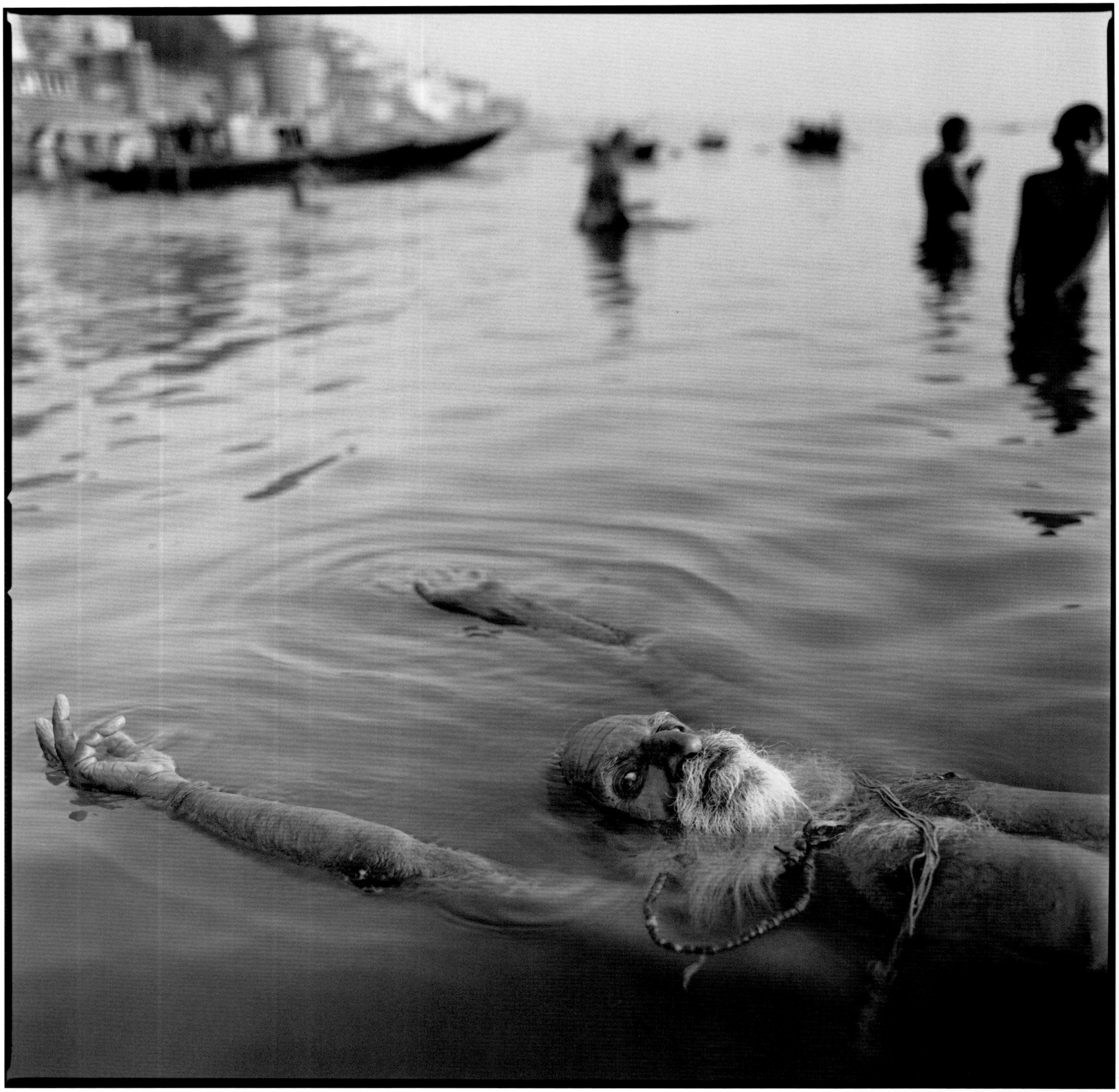

PLATE 33

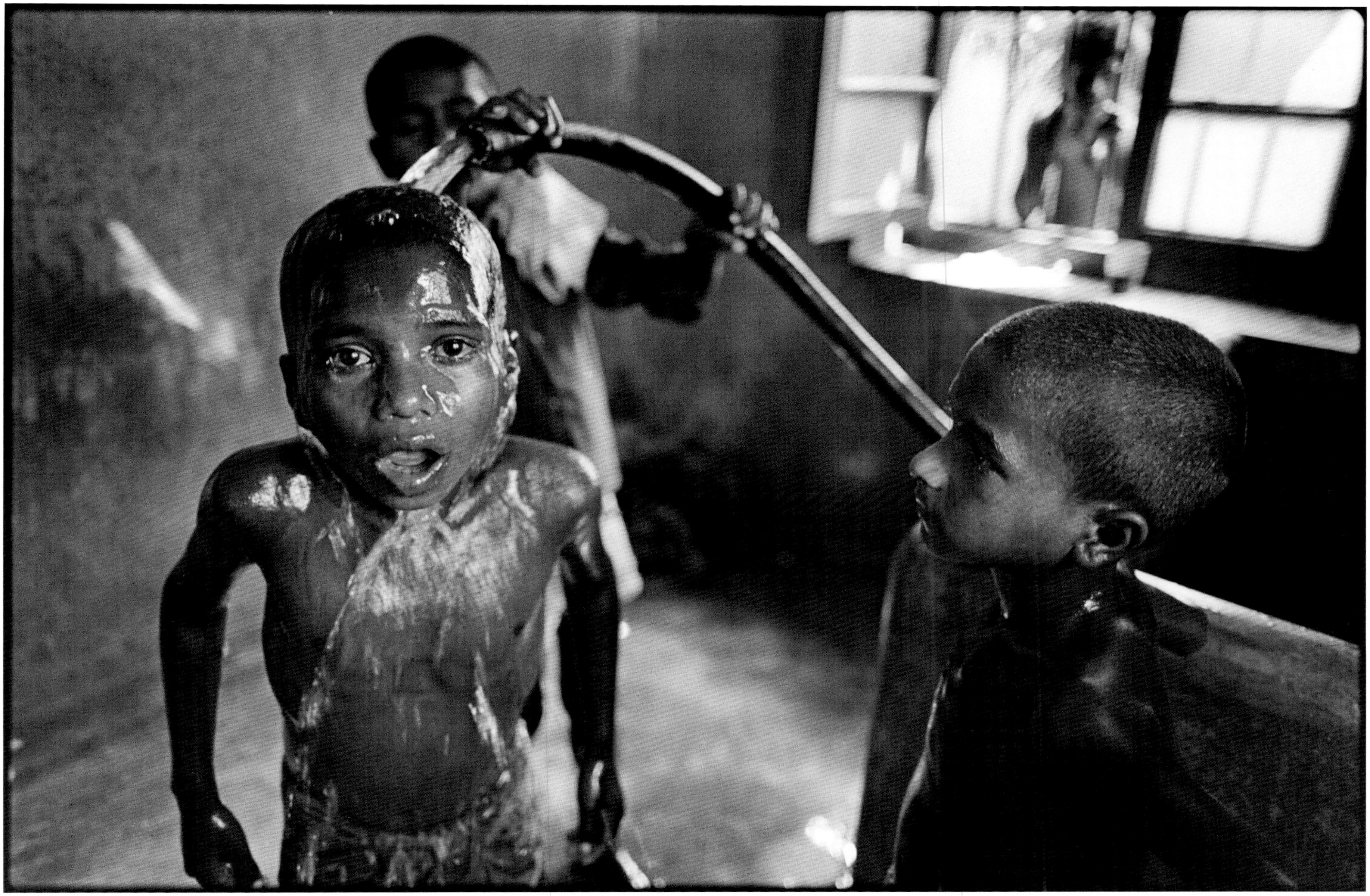

PLATE 34

PLATE 35

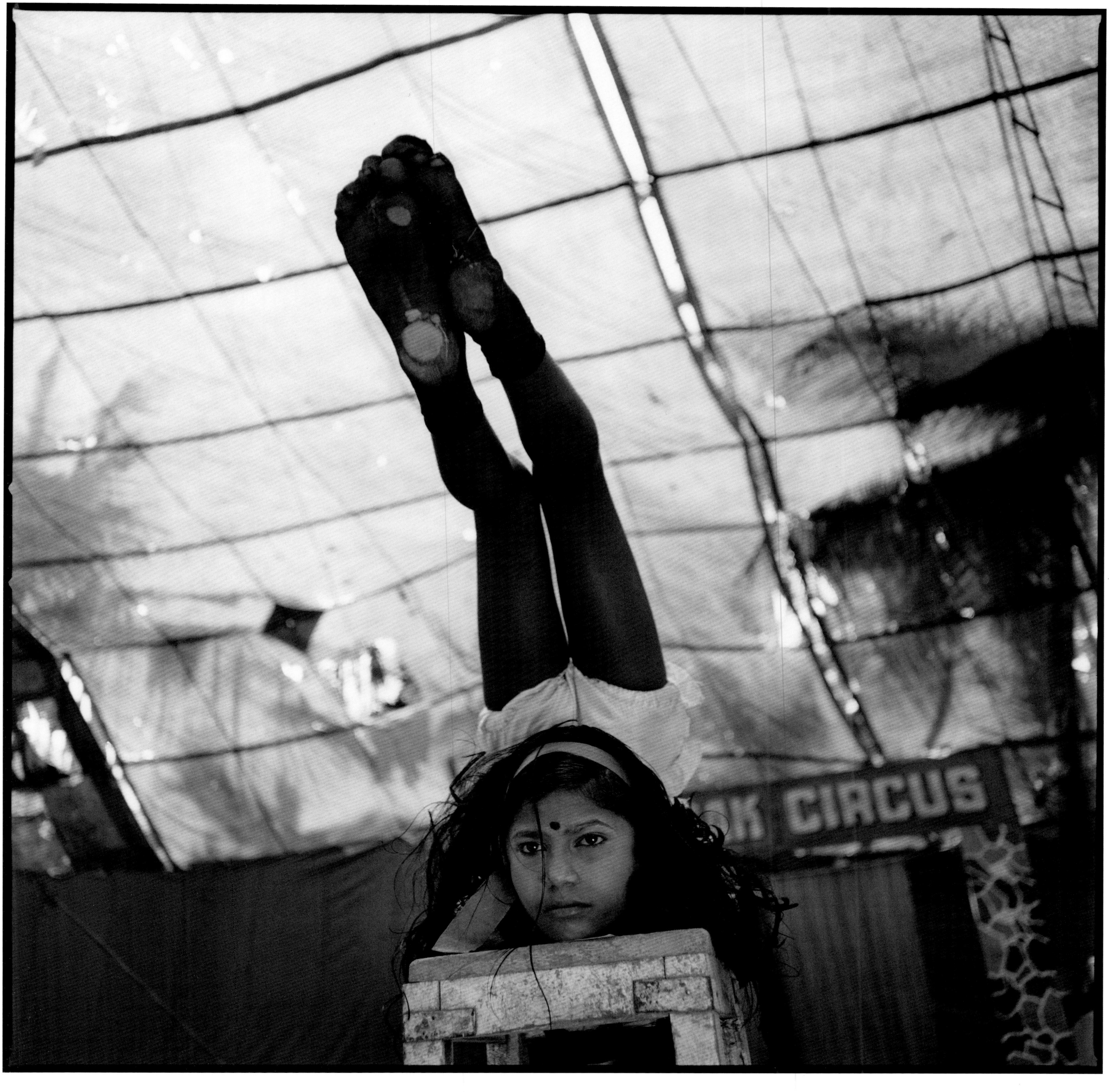
CIRCUS

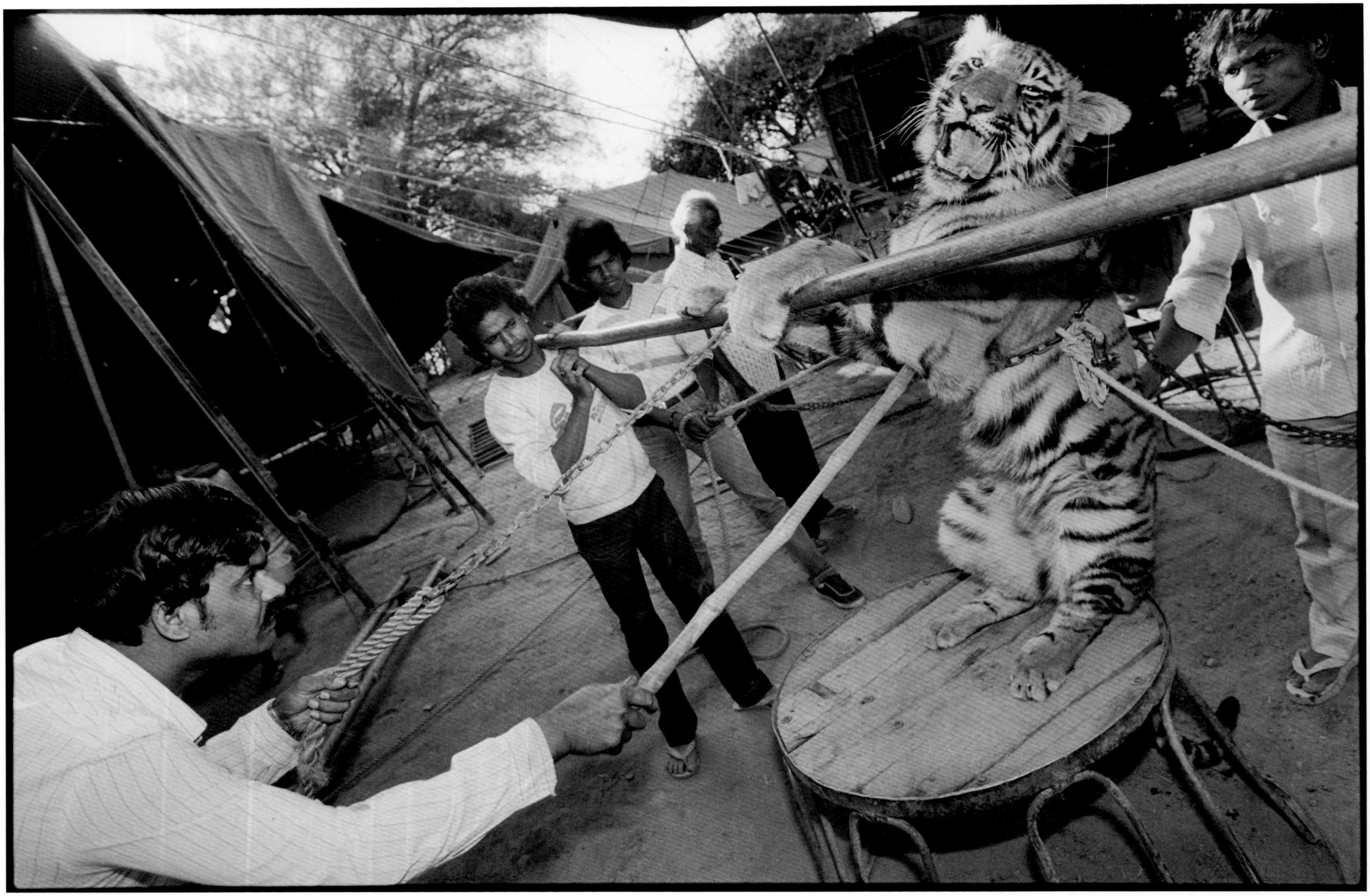

PLATE 36

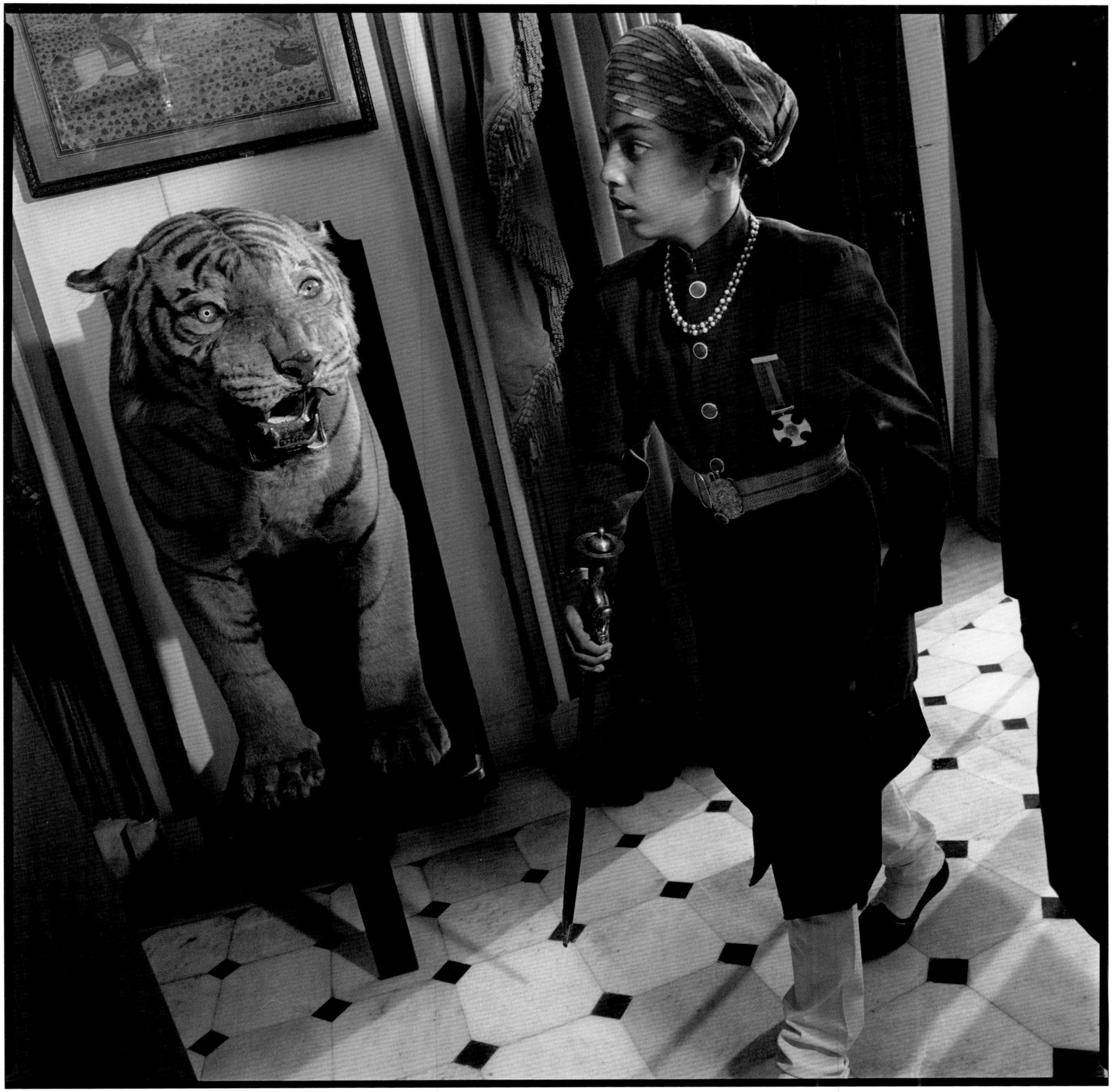

PLATE 37

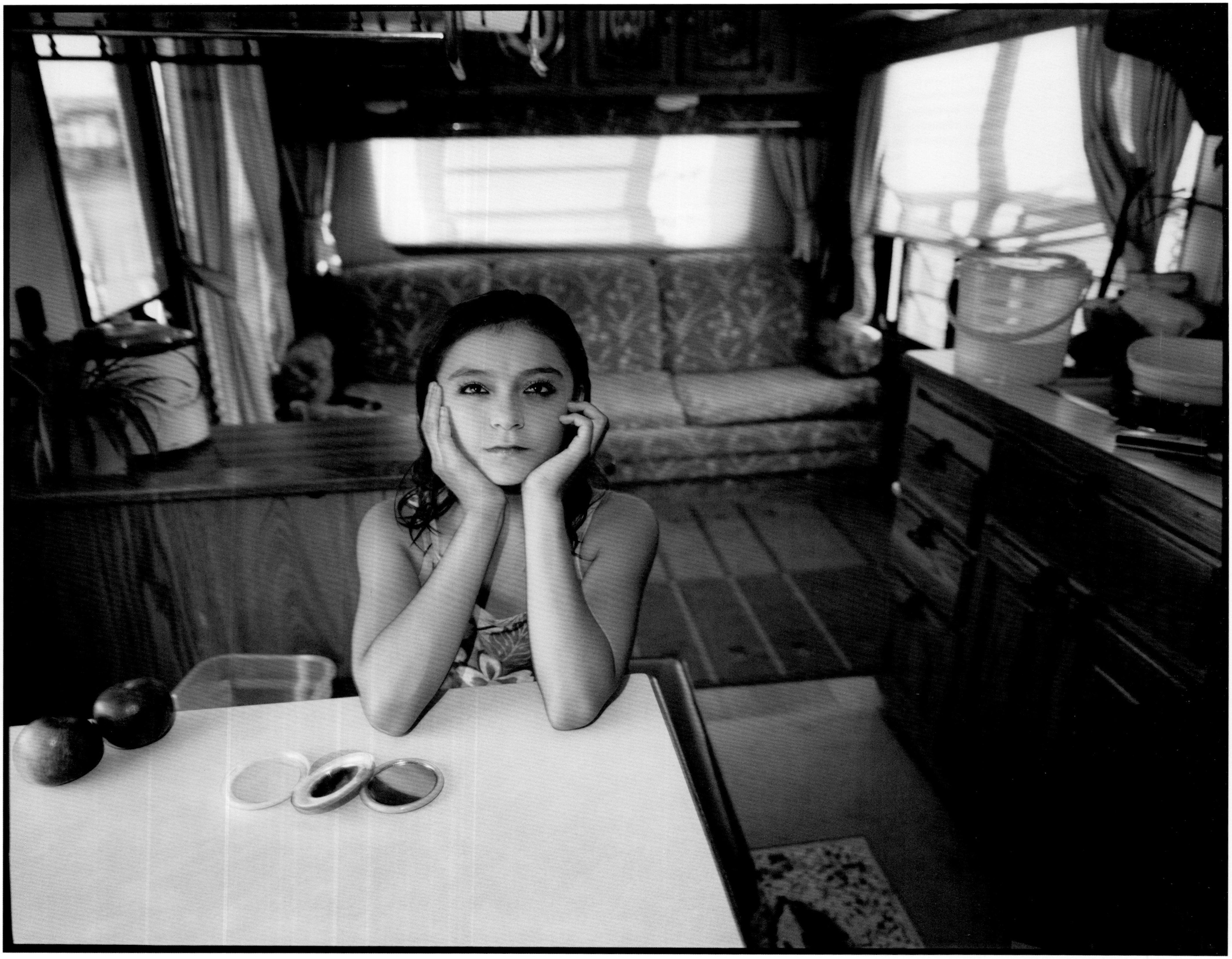

PLATE 38

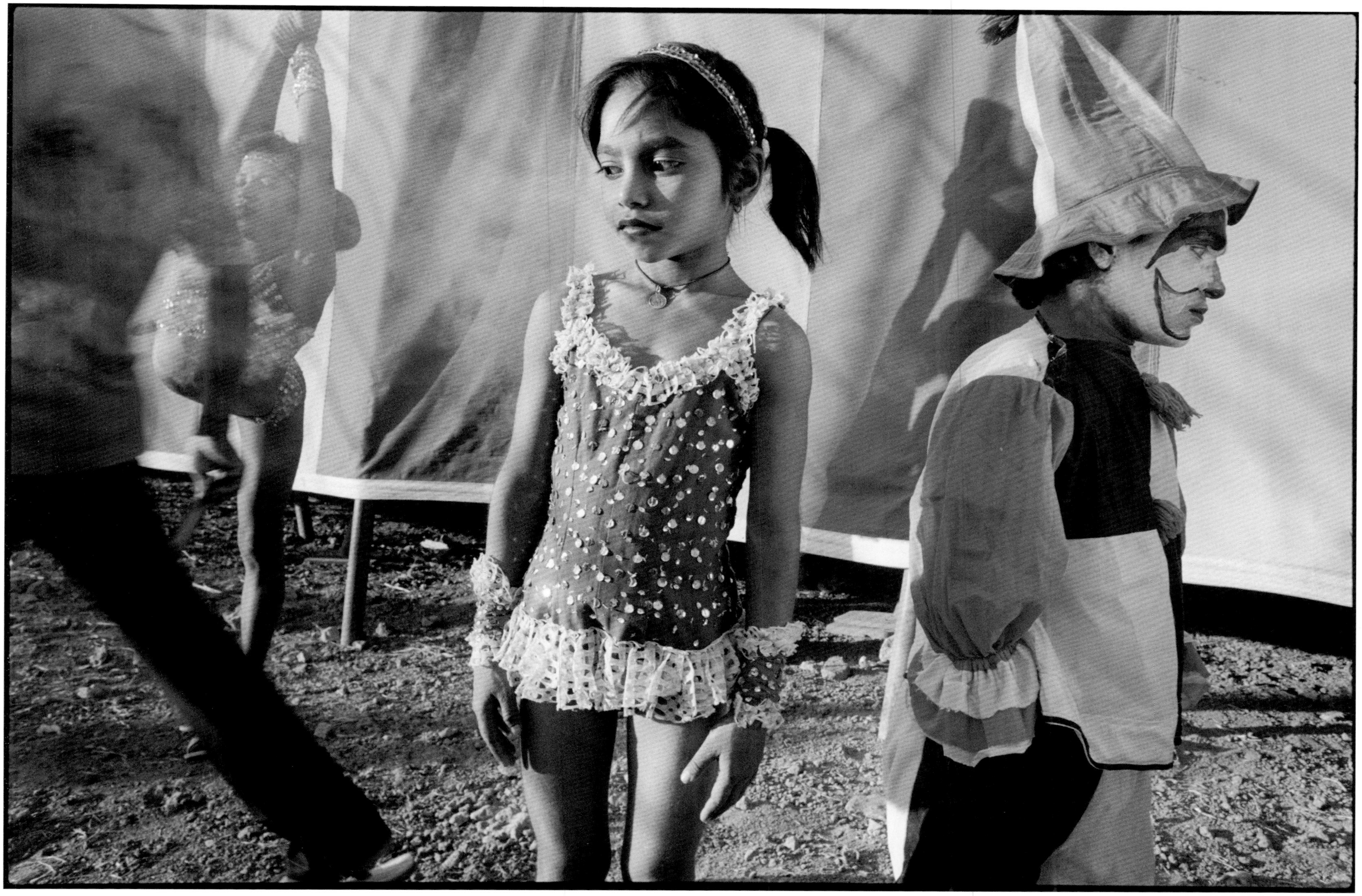

PLATE 39

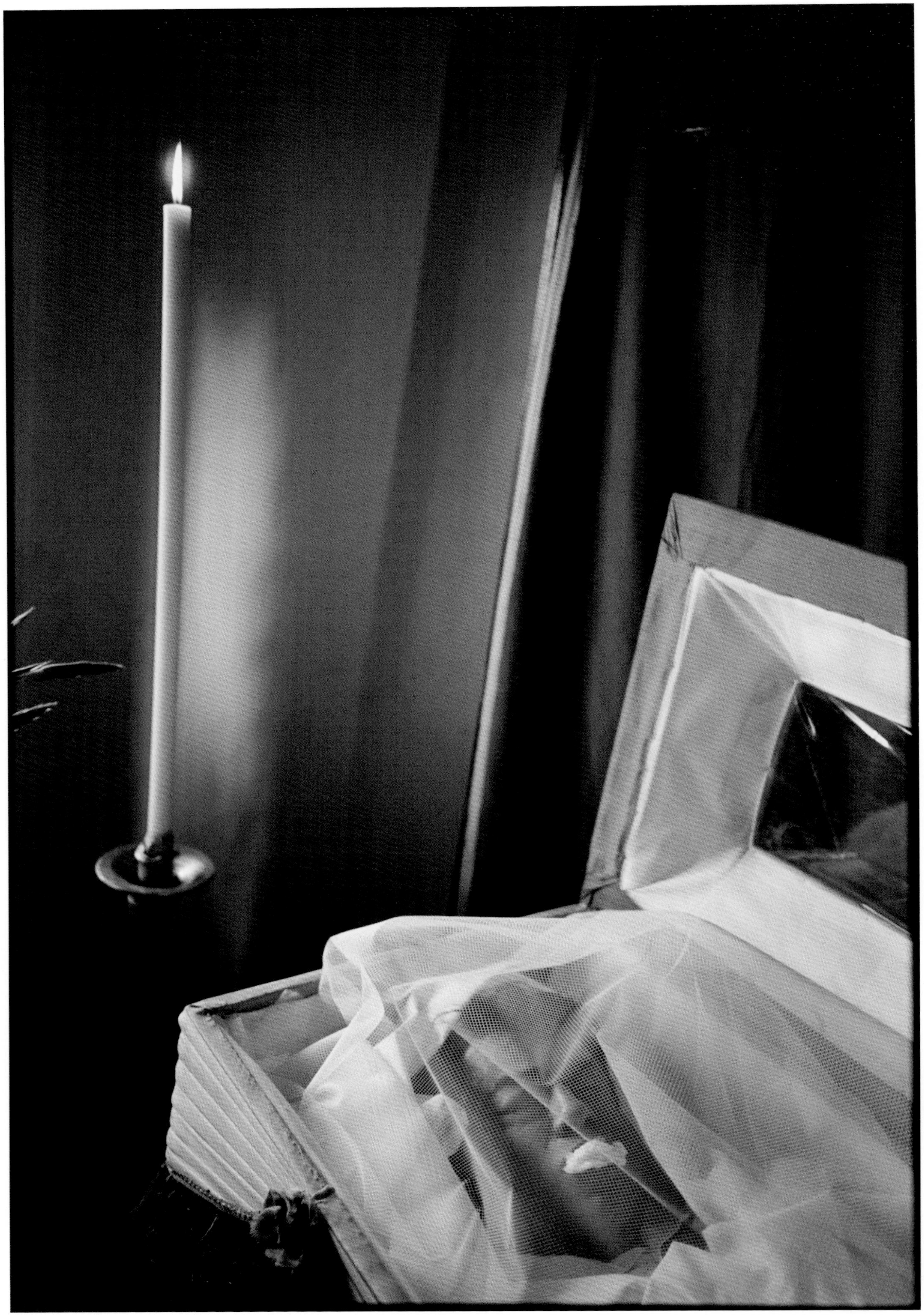

PLATE 40

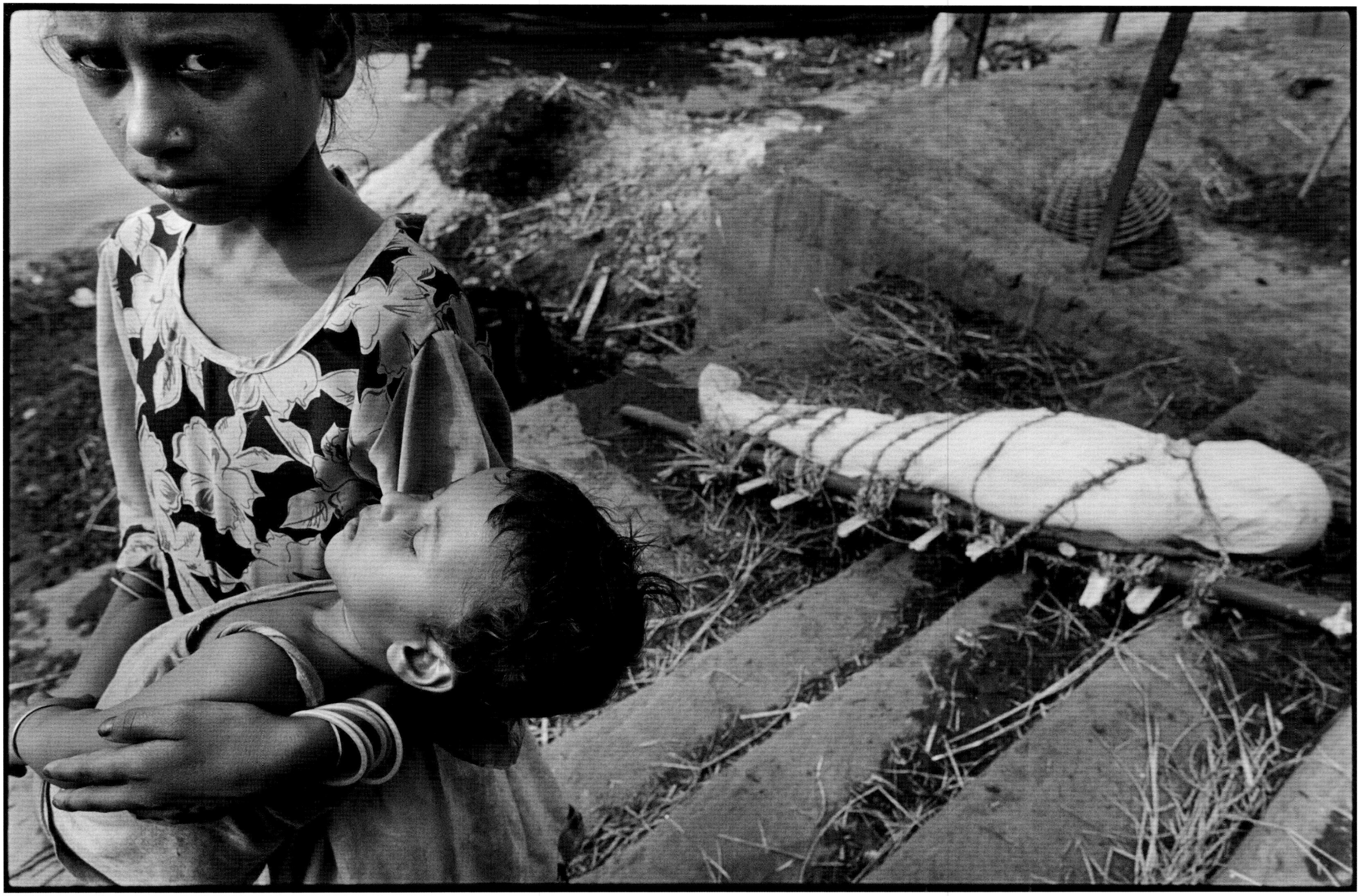

PLATE 41

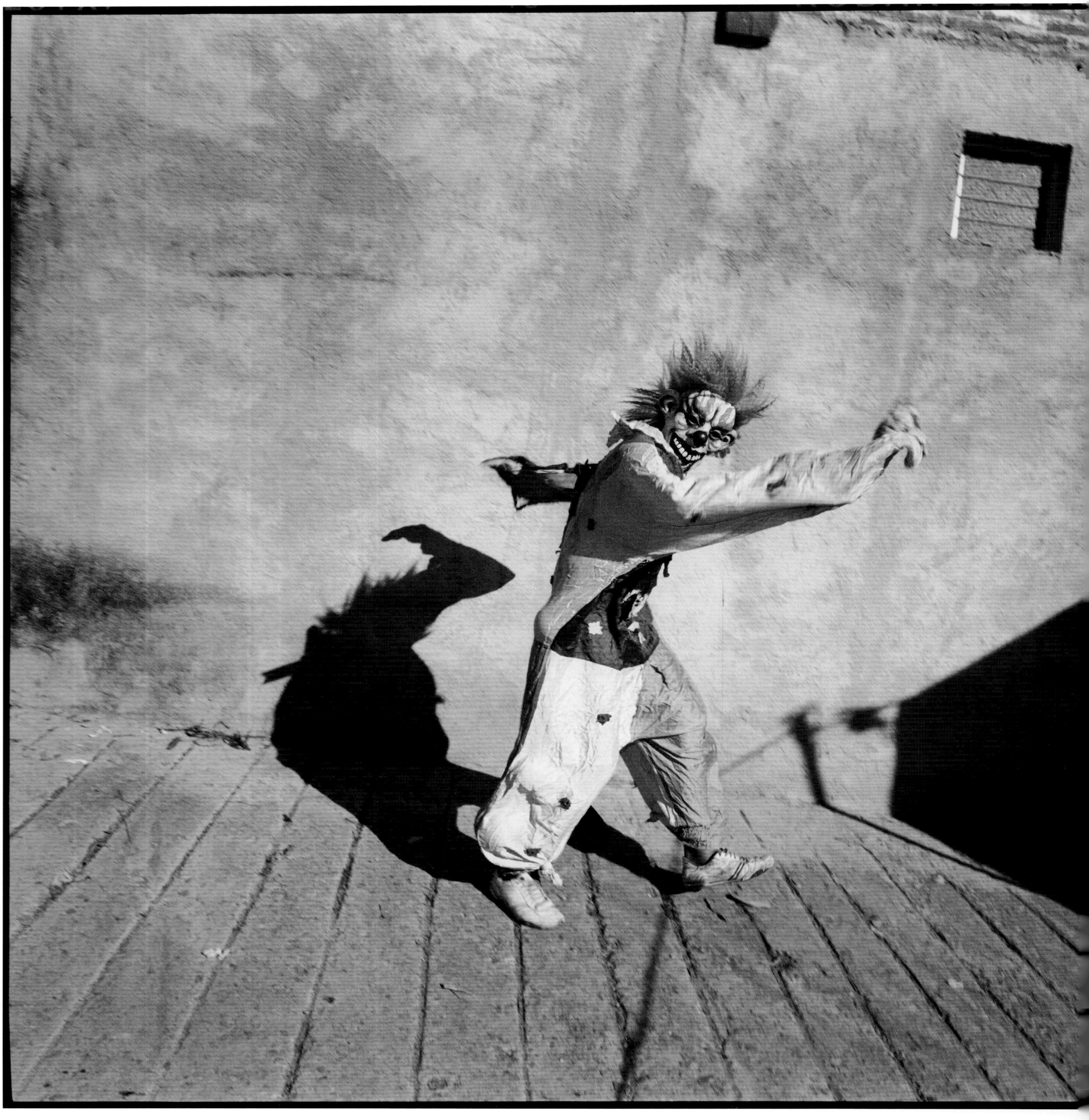

PLATE 42

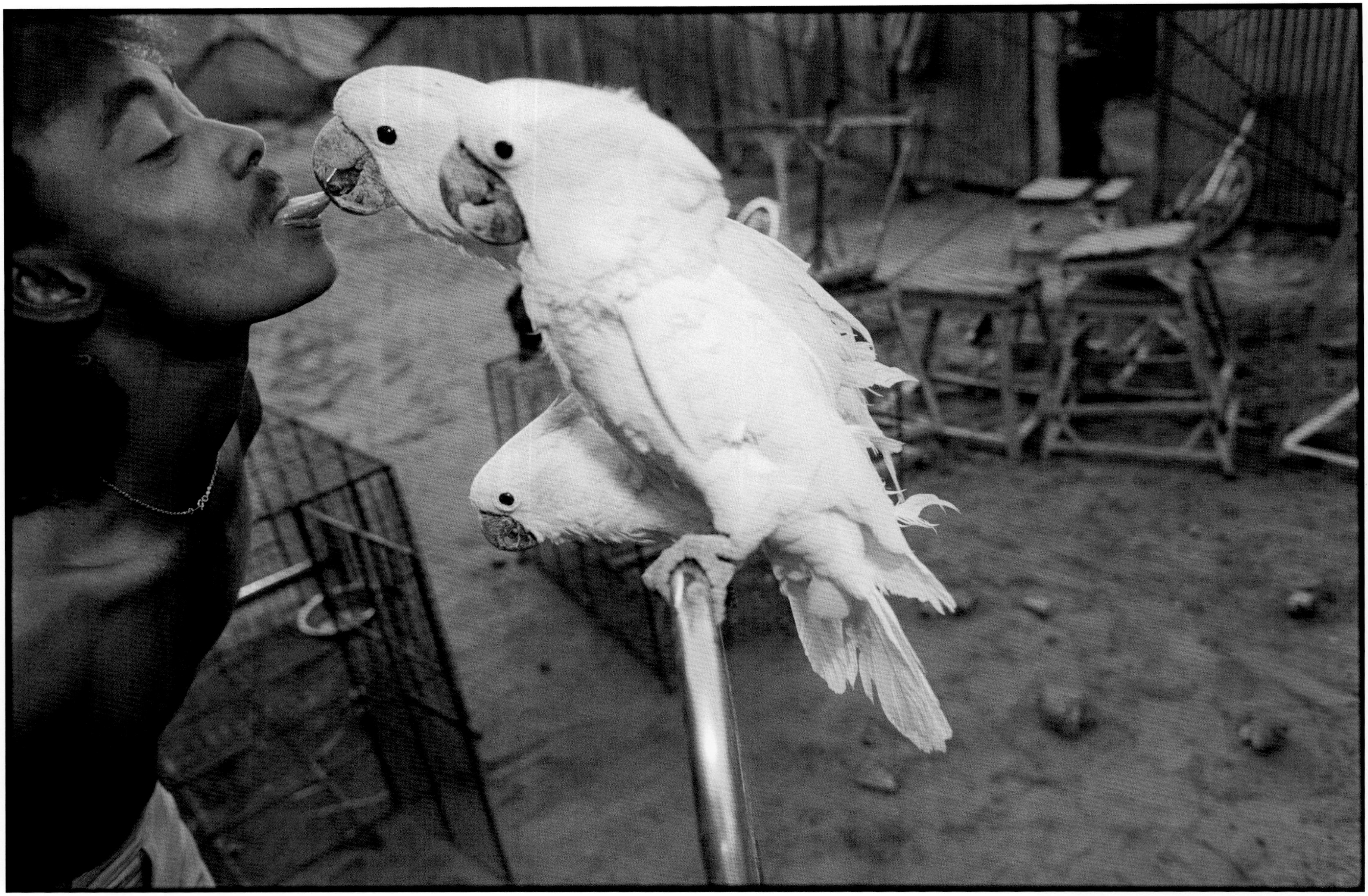

PLATE 43

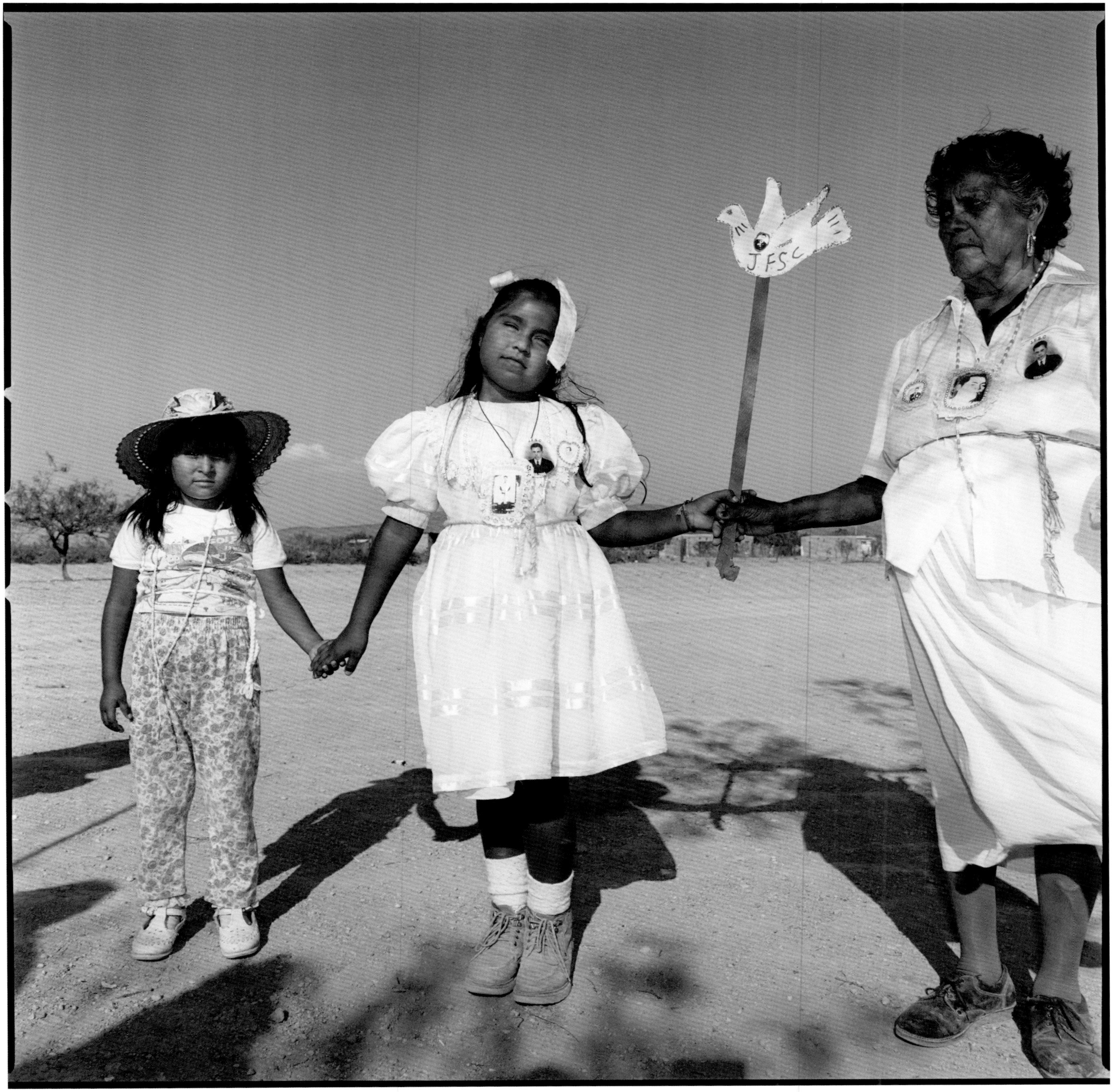

PLATE 44

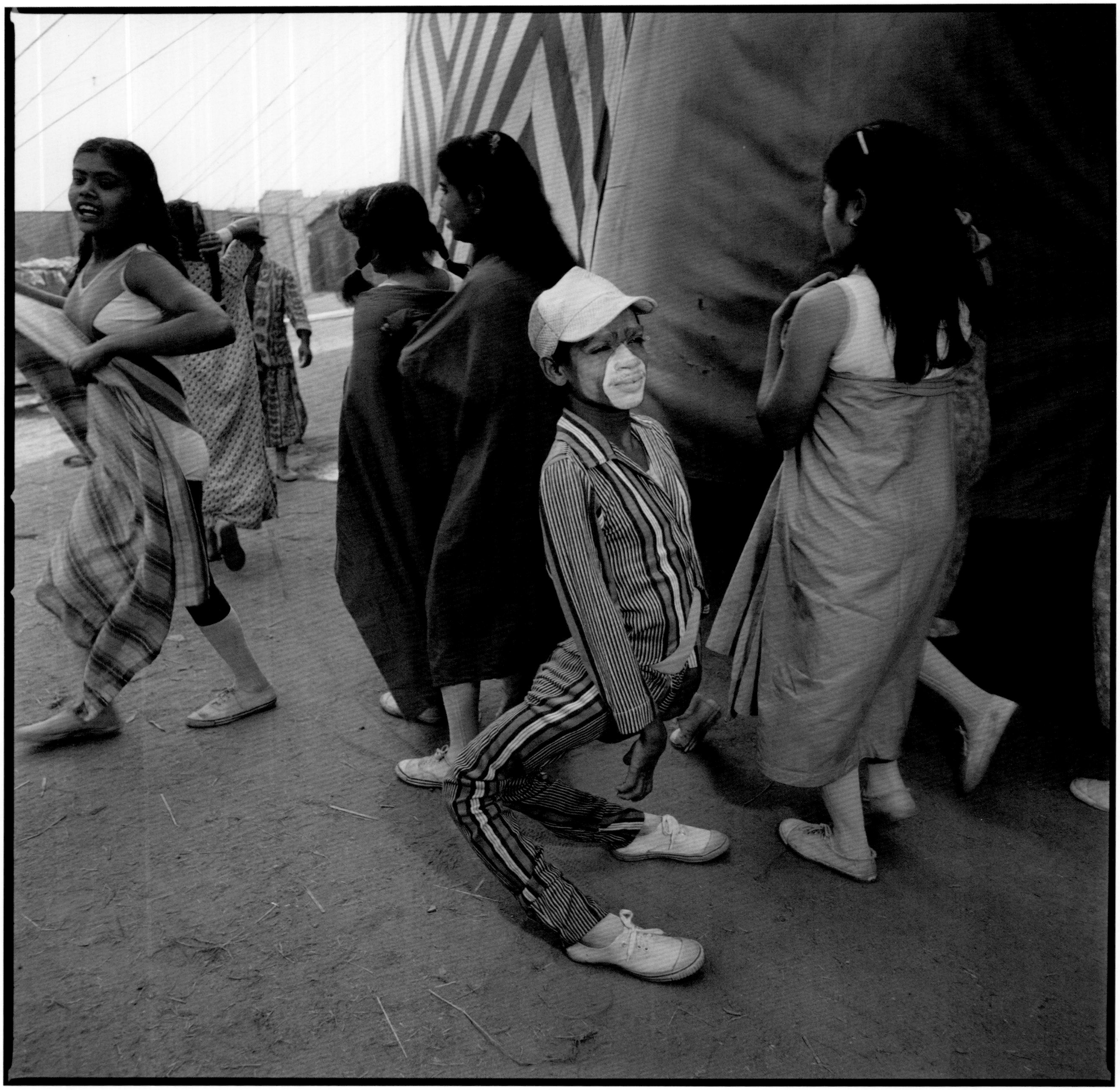

PLATE 45

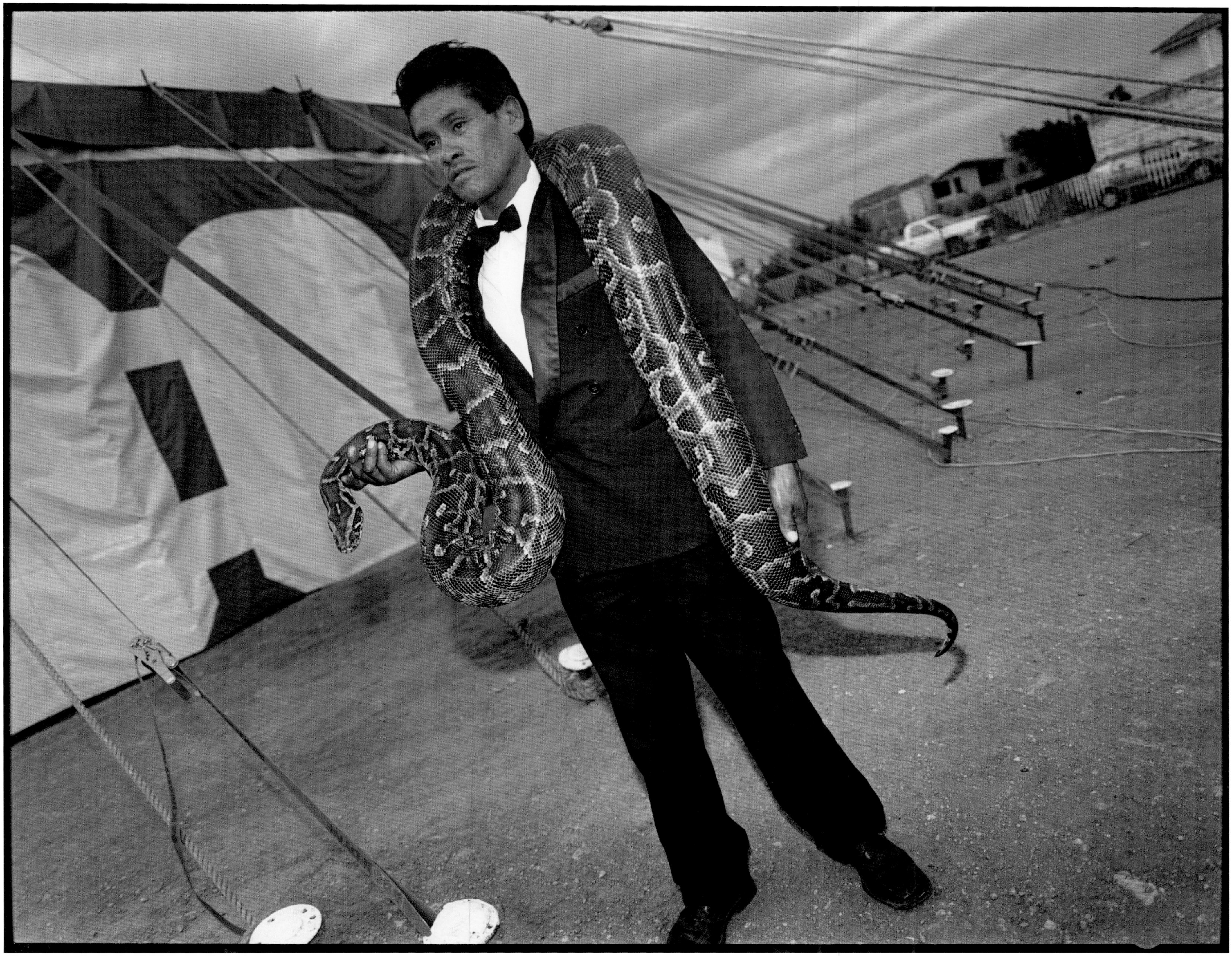

PLATE 46

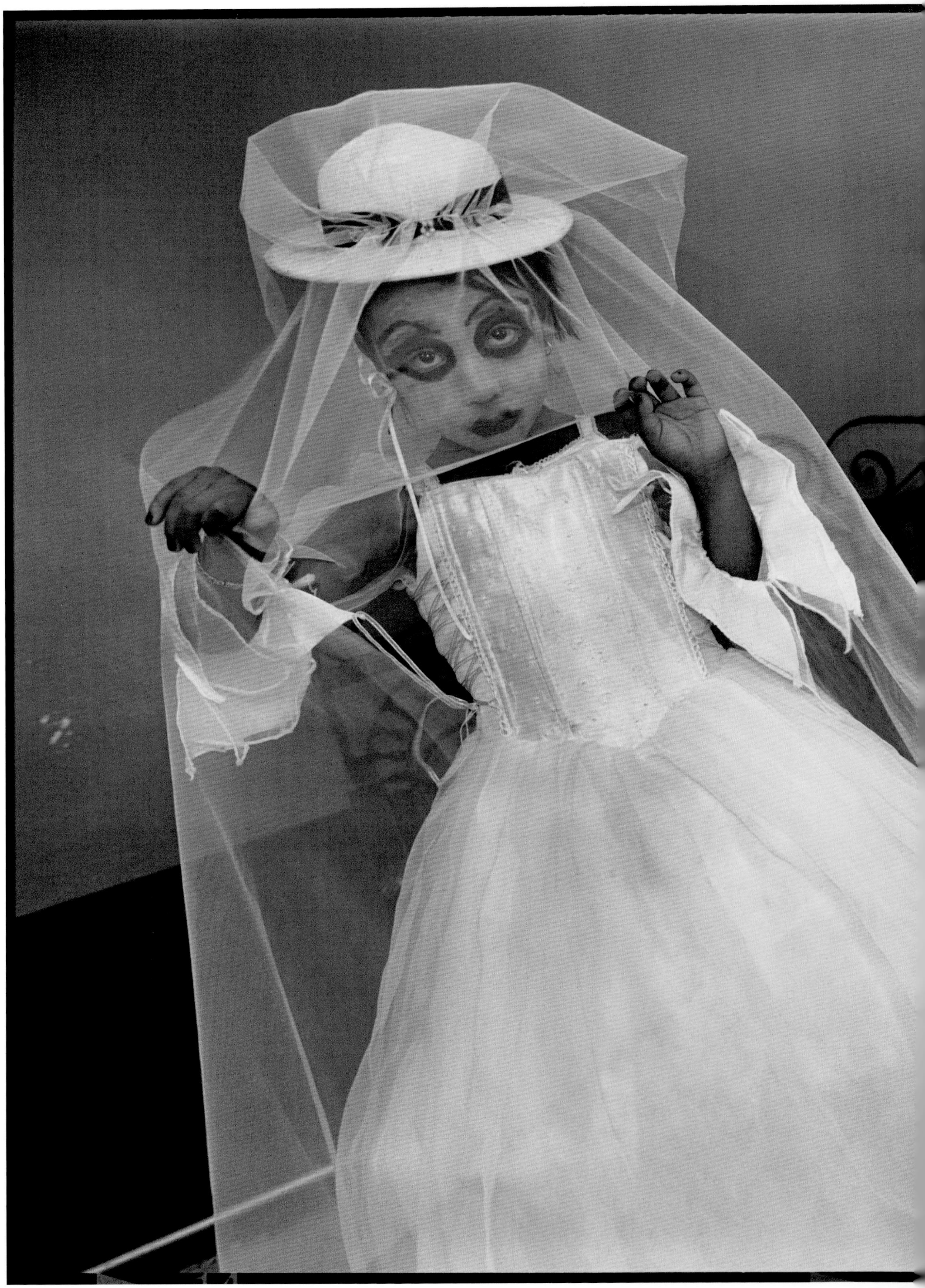

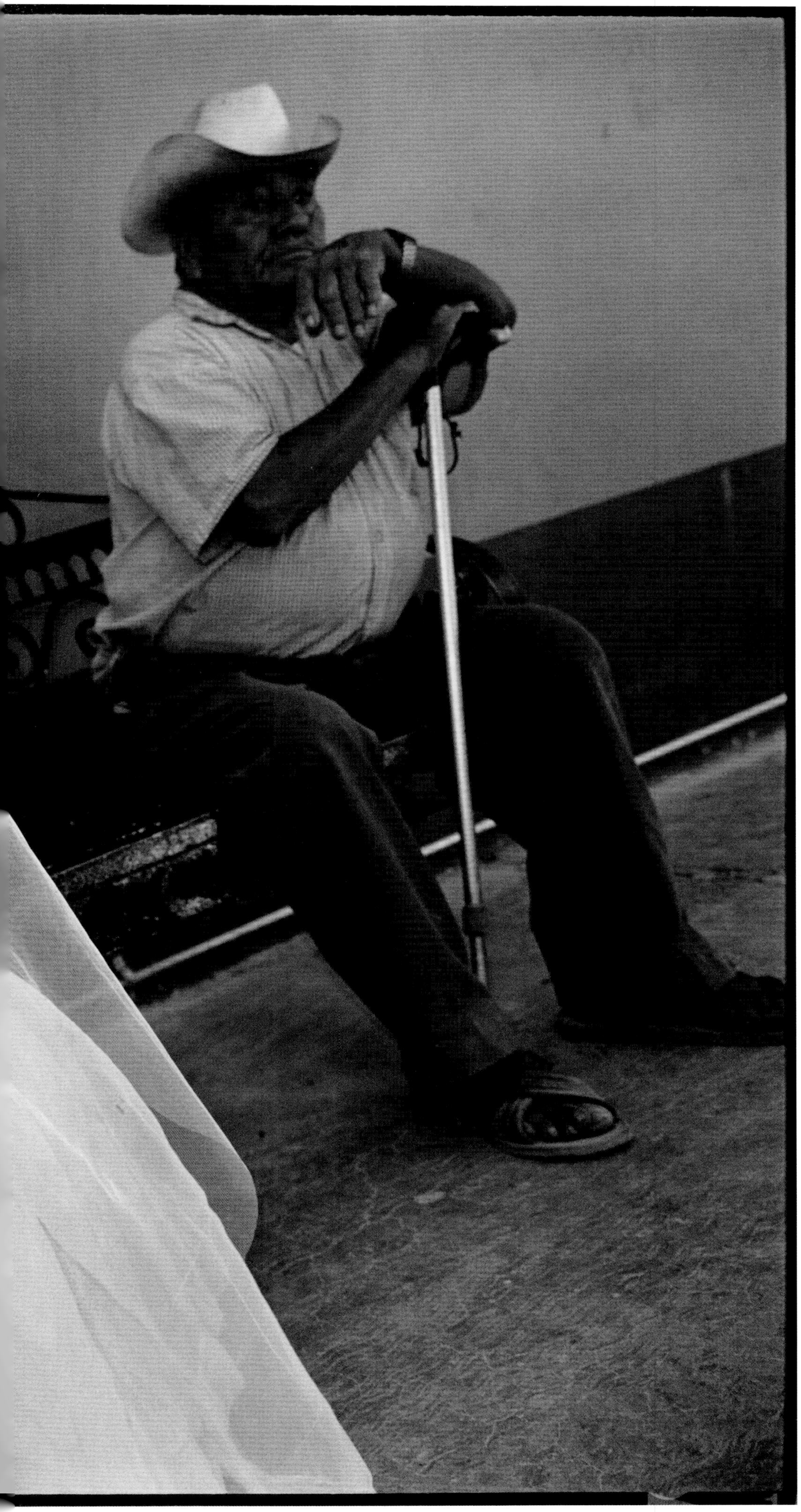

PLATE 47

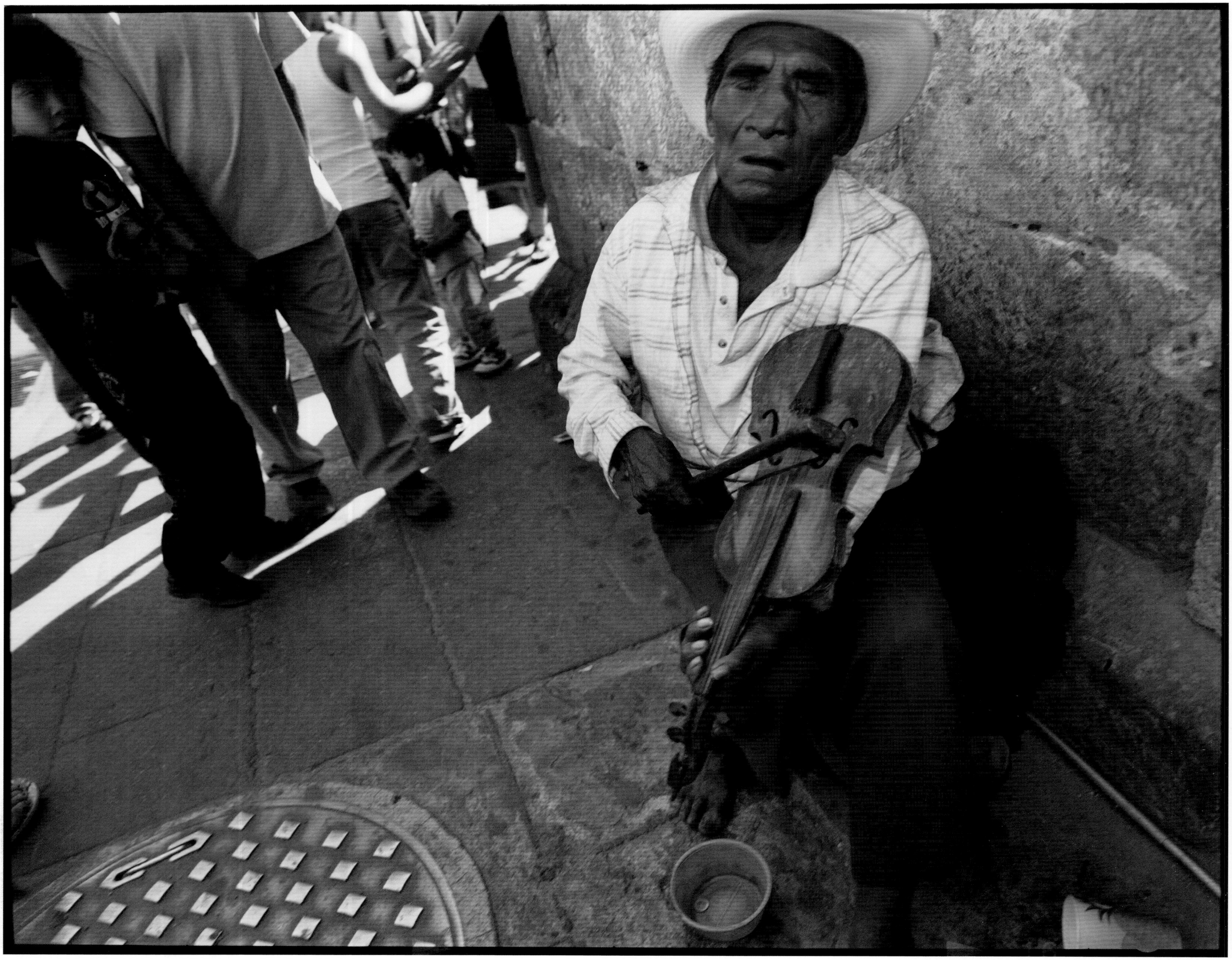

PLATE 48

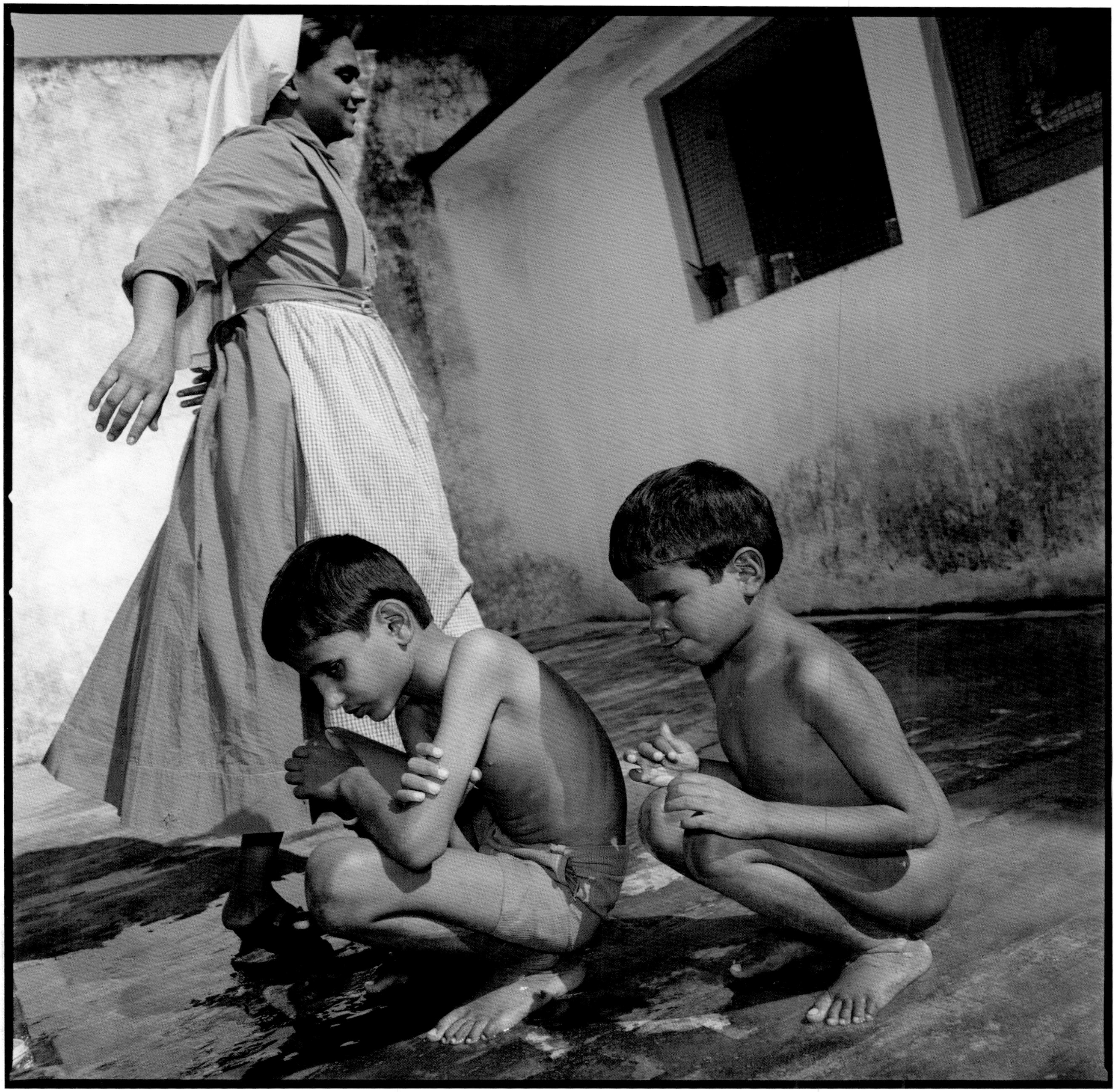

PLATE 49

PLATE 50

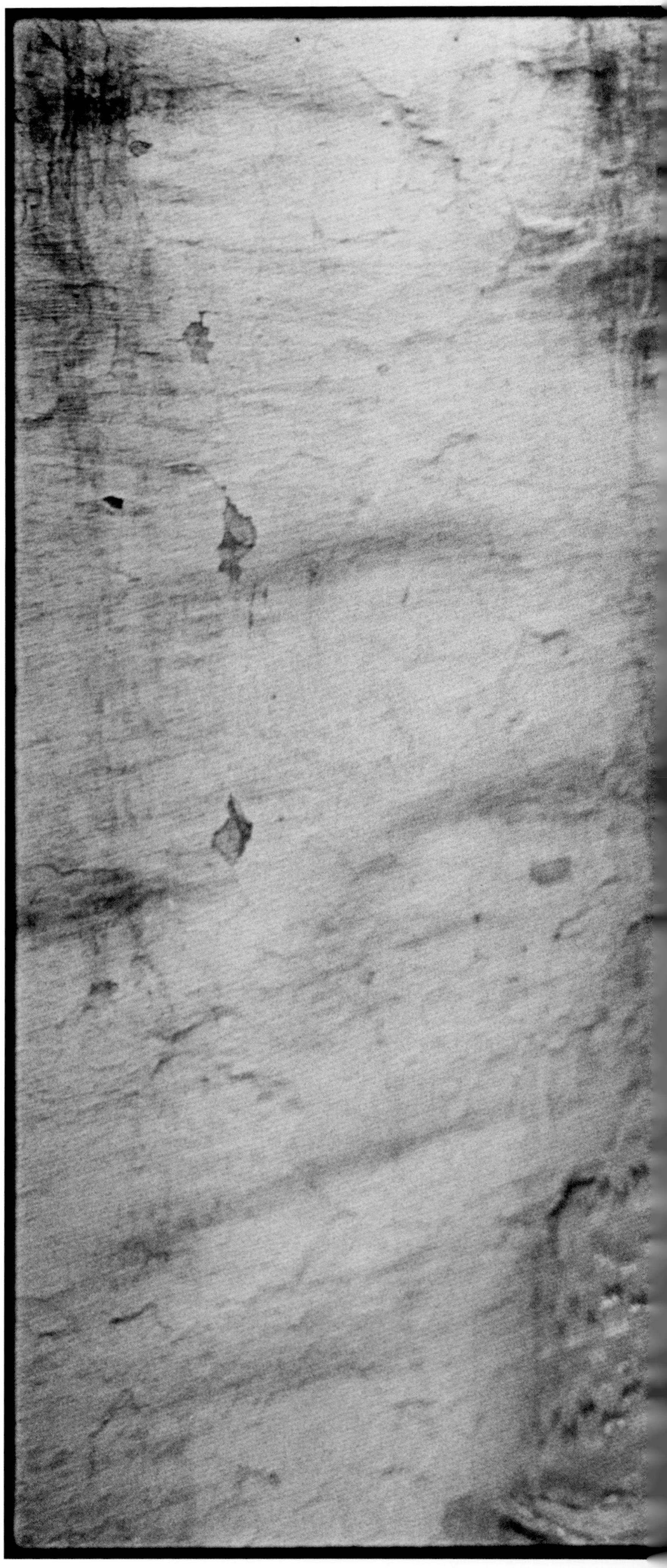

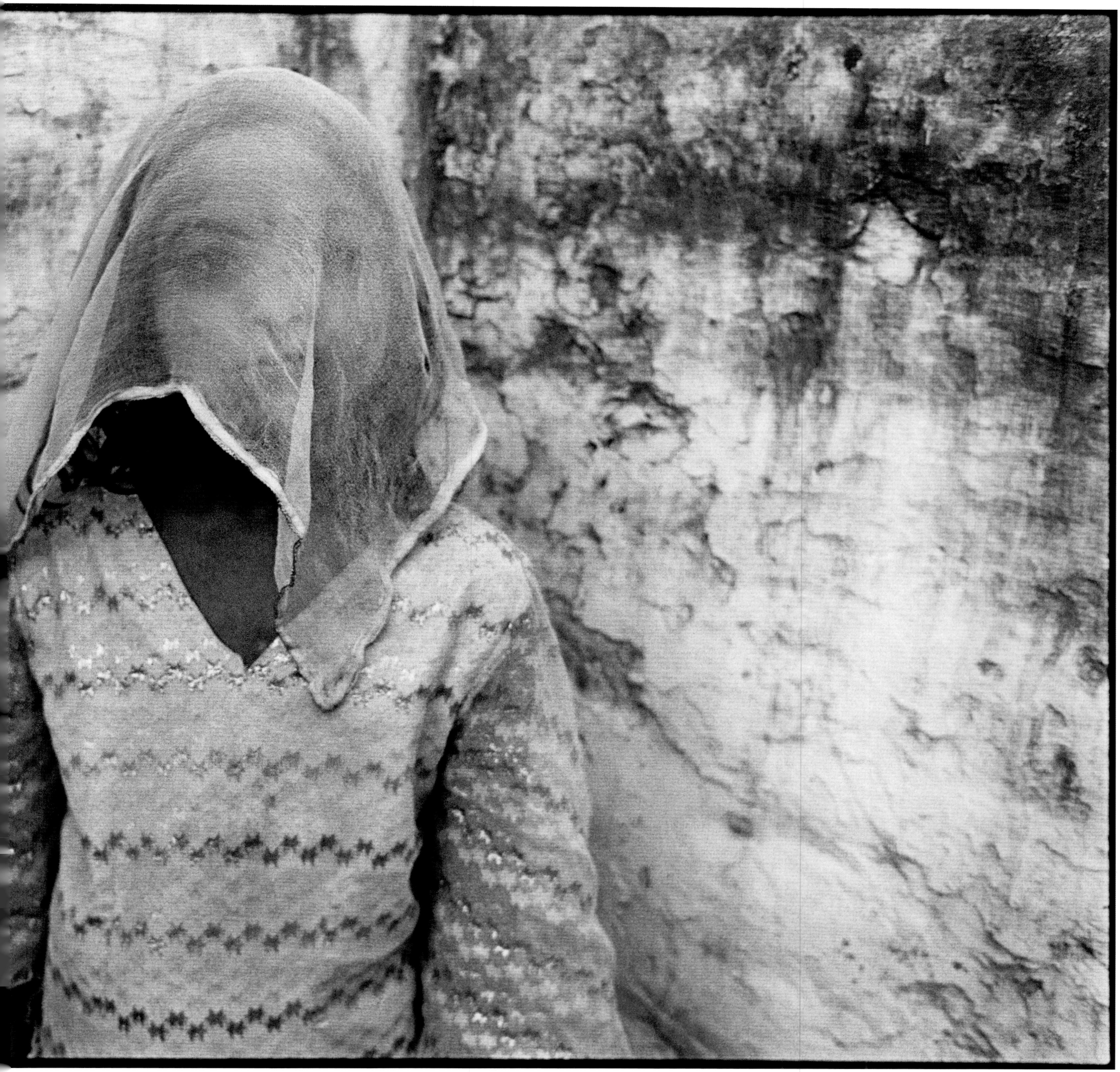

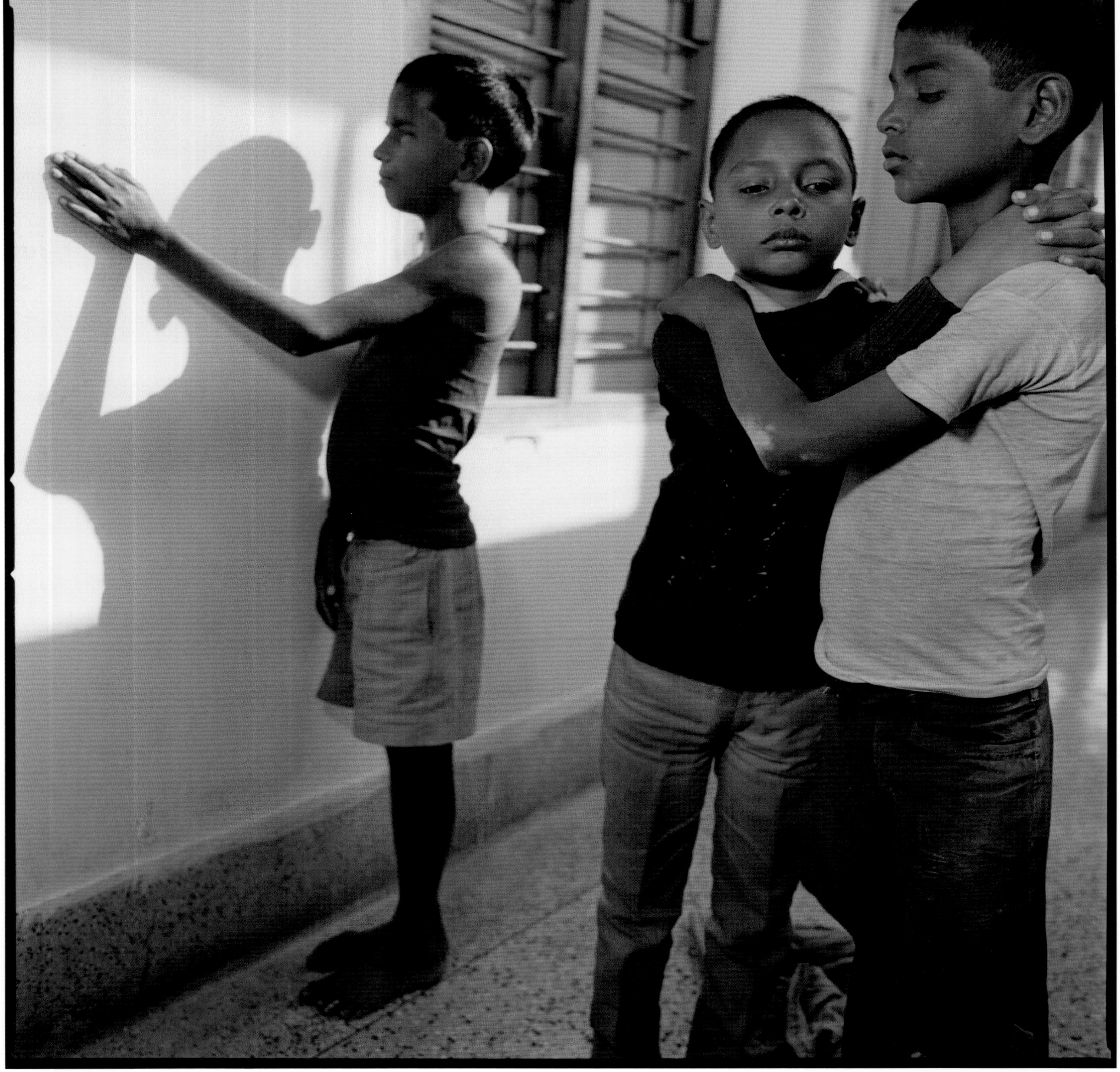

PLATE 51

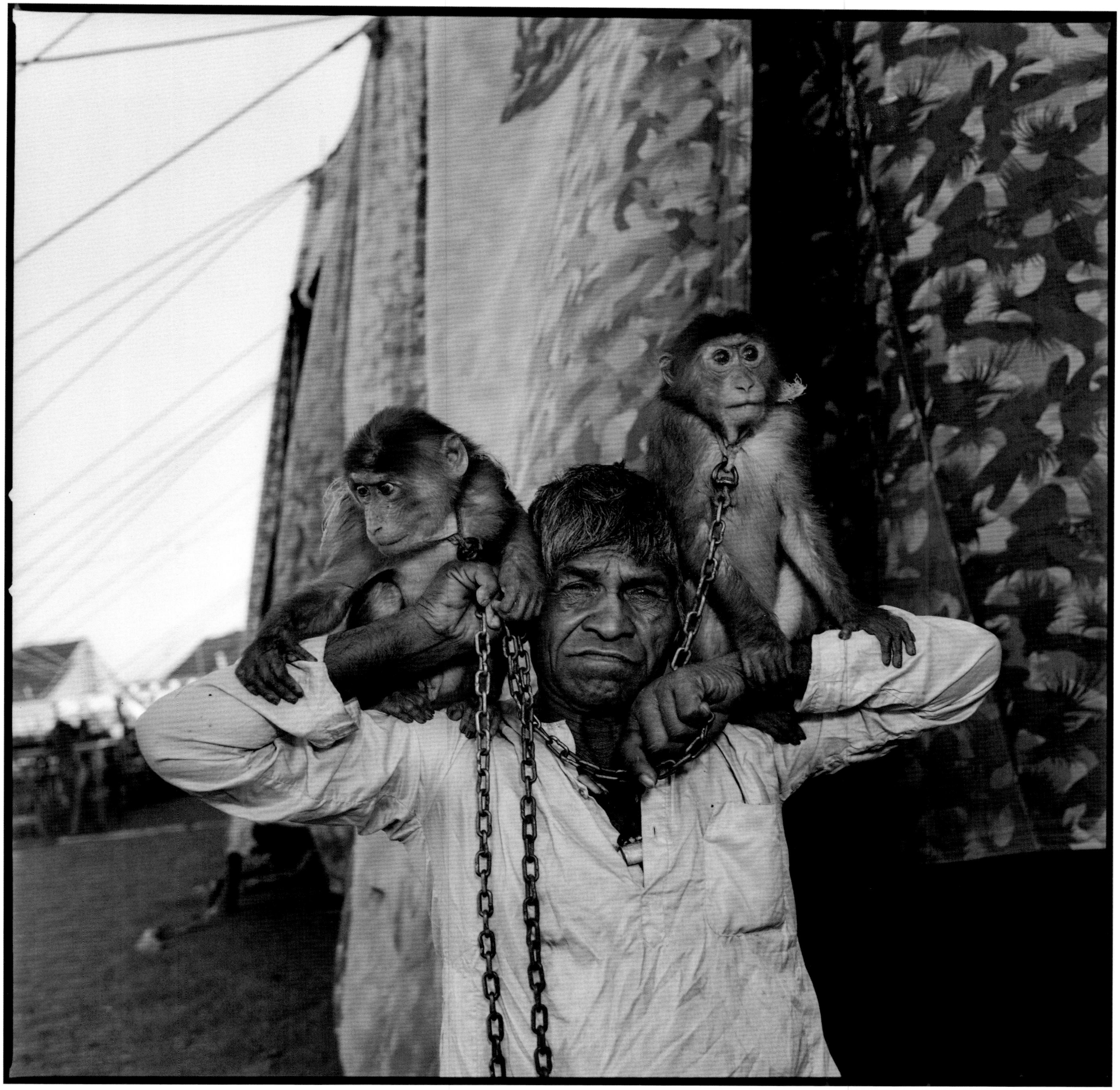

PLATE 52

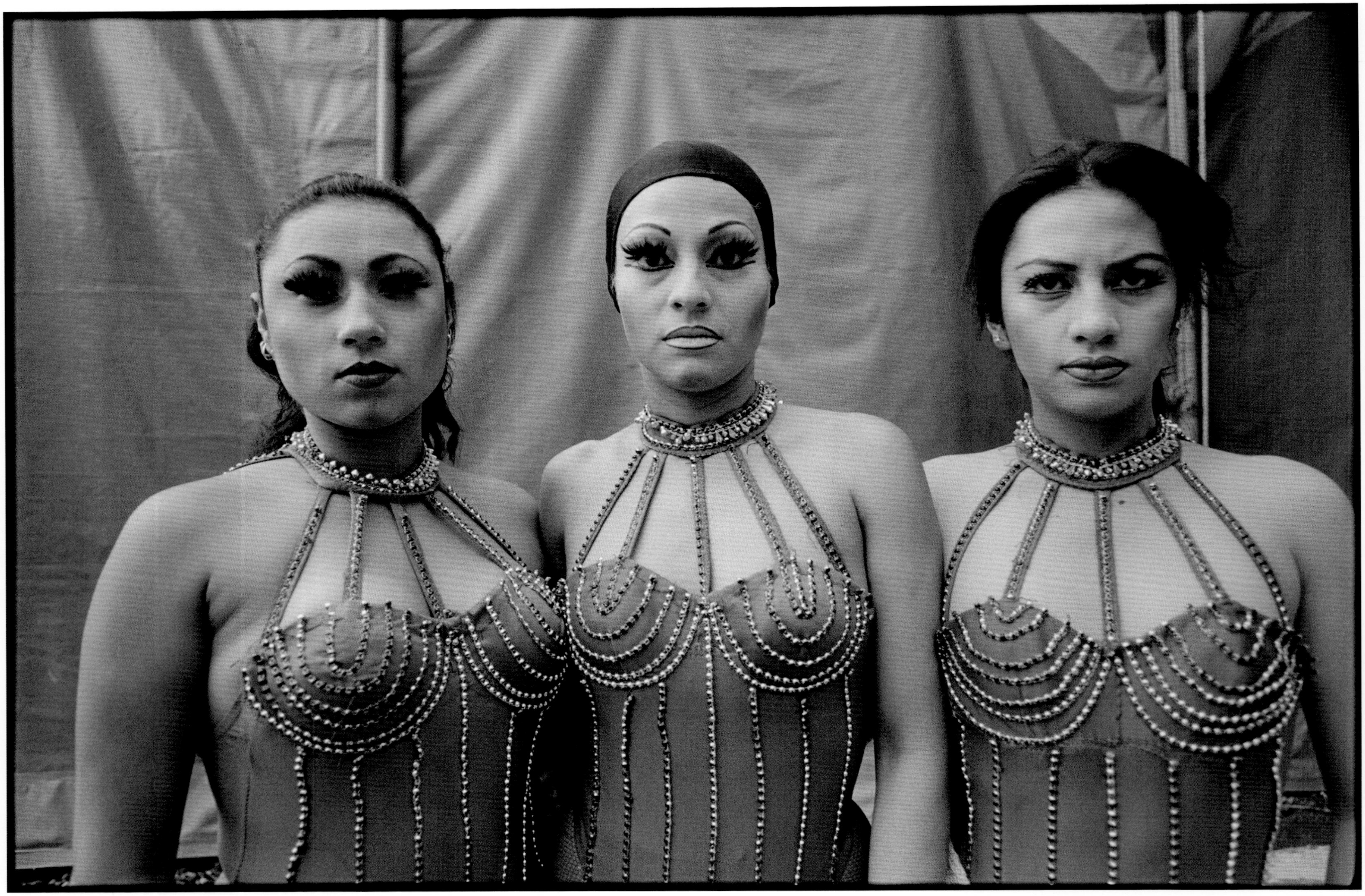

PLATE 53

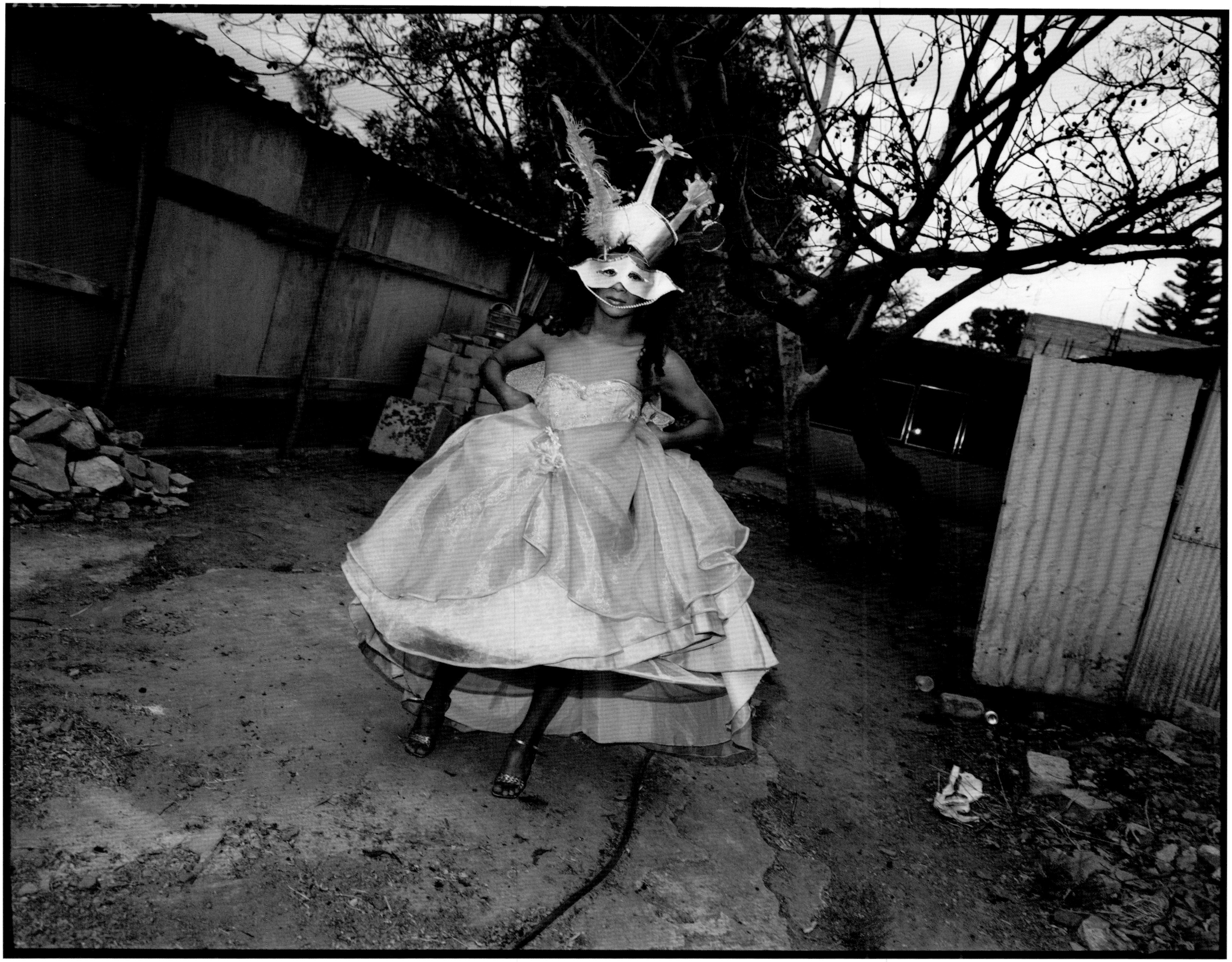

PLATE 54

PLATE 55

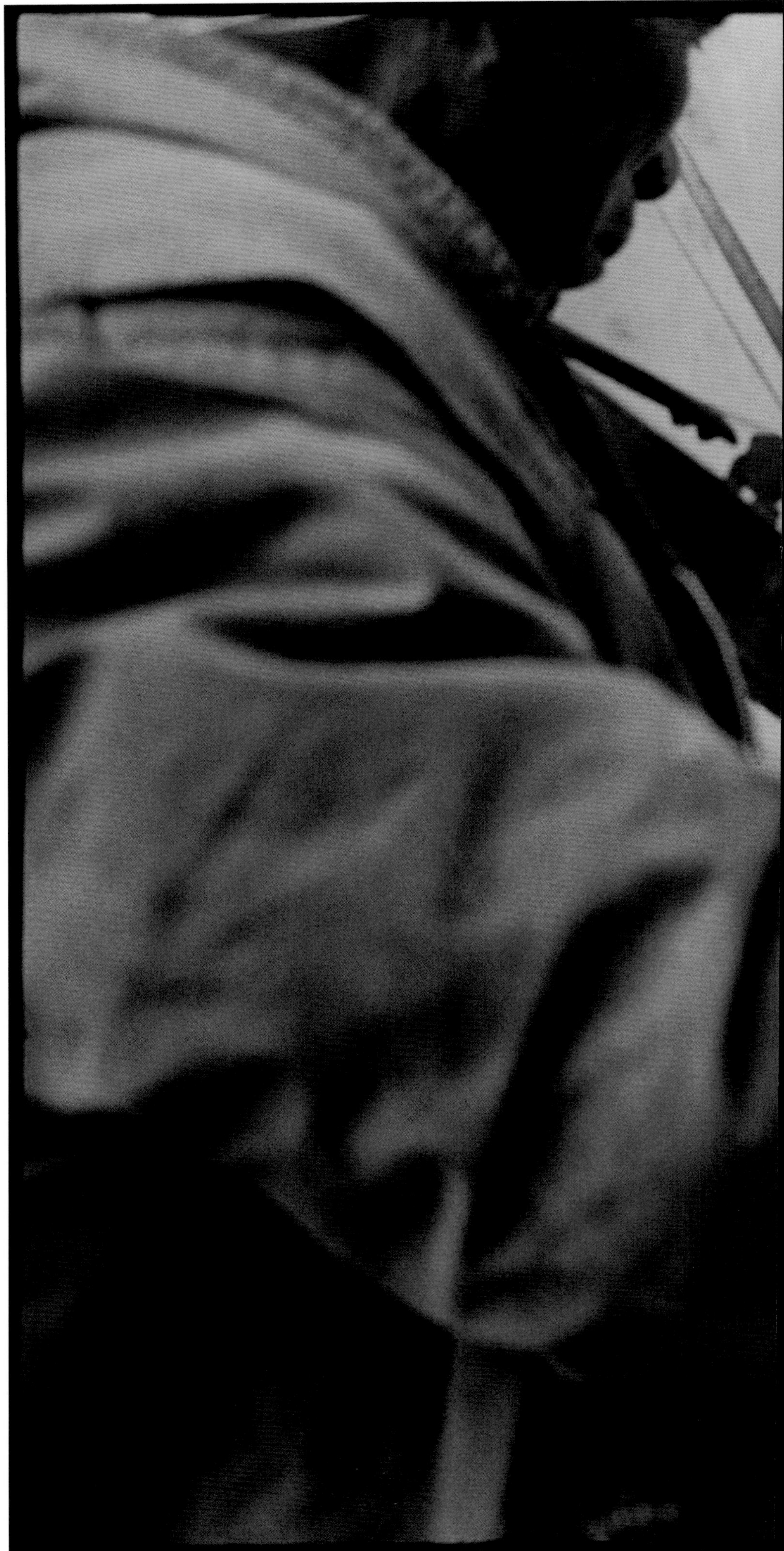

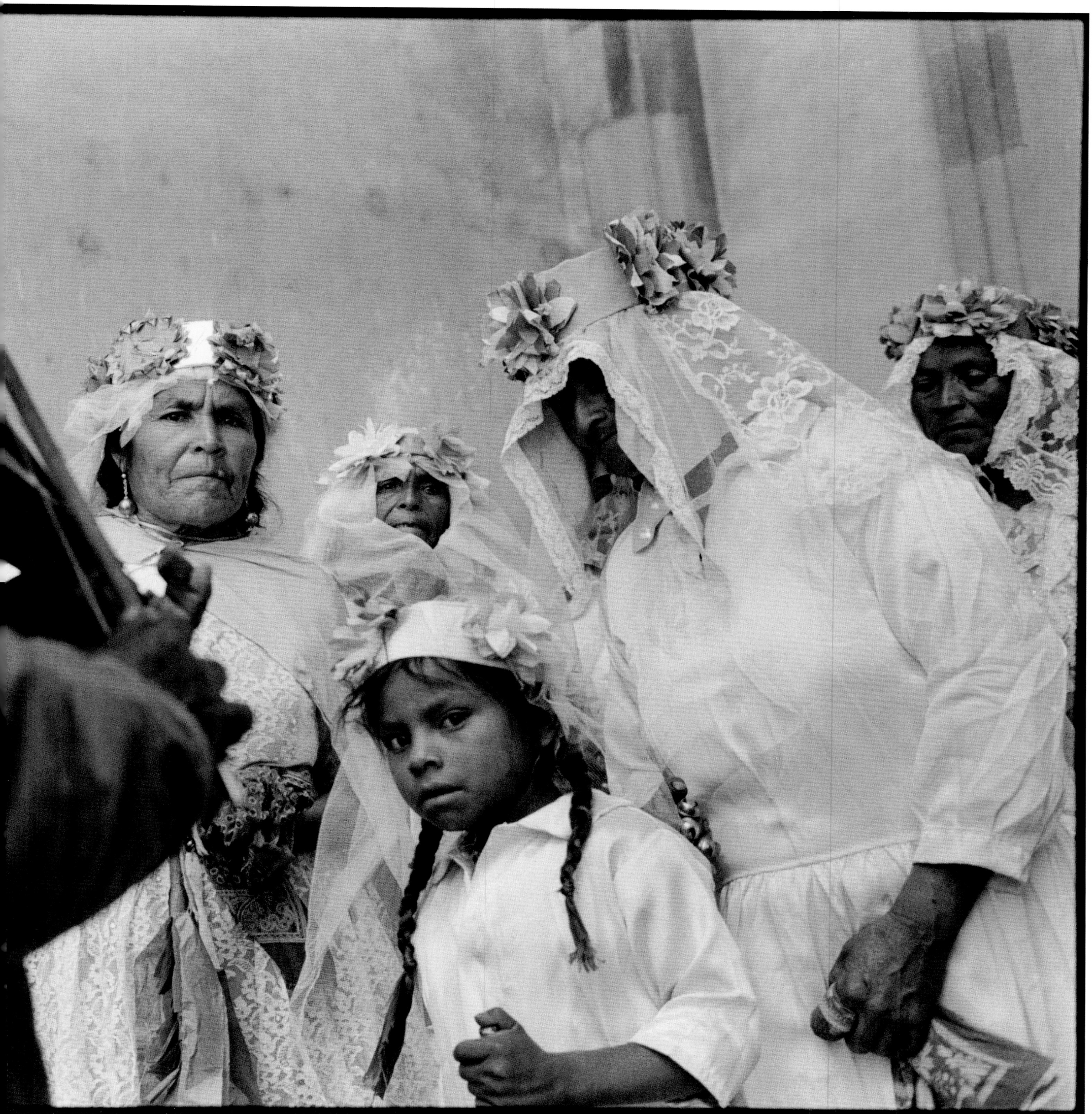

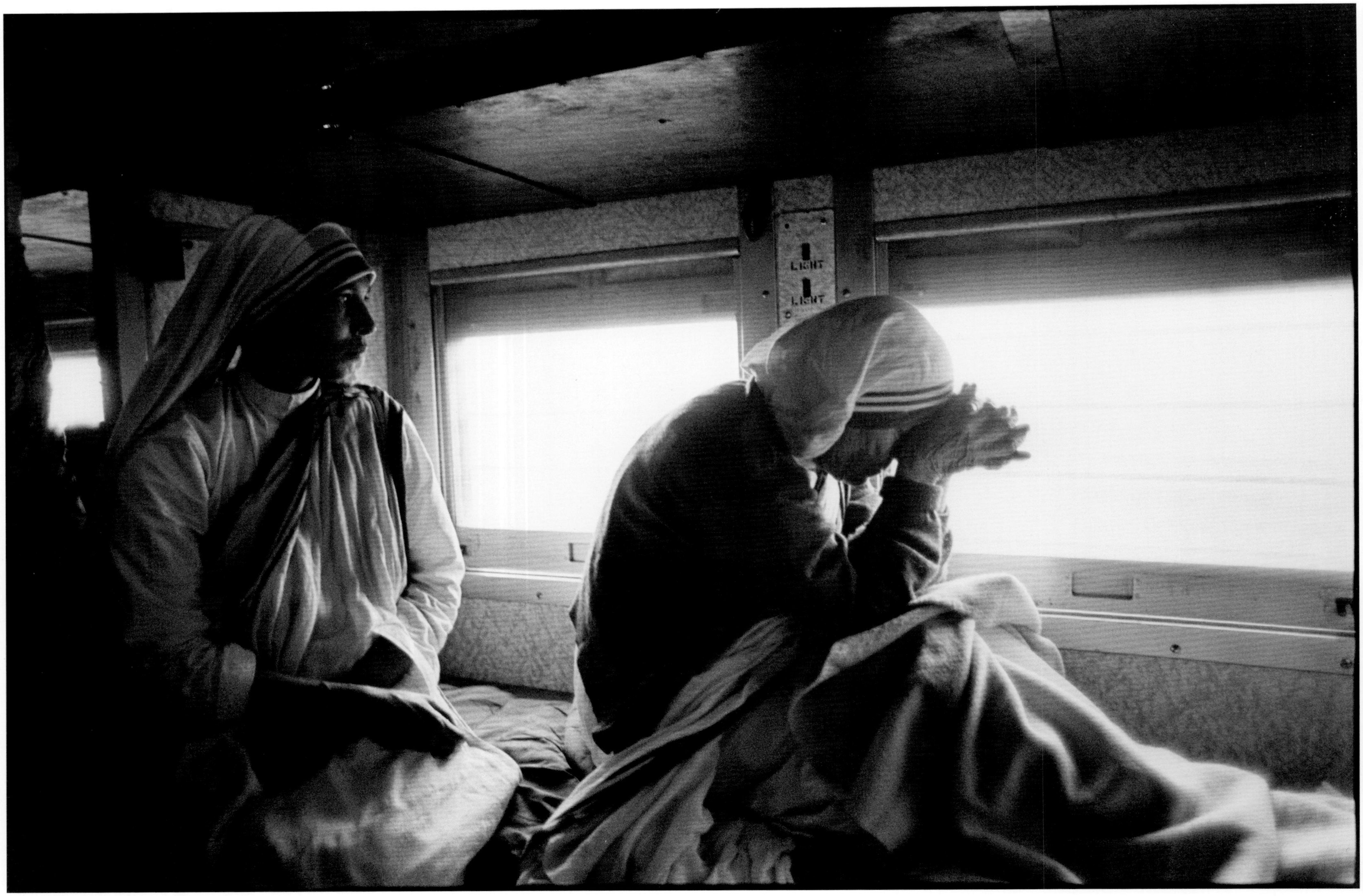

PLATE 56

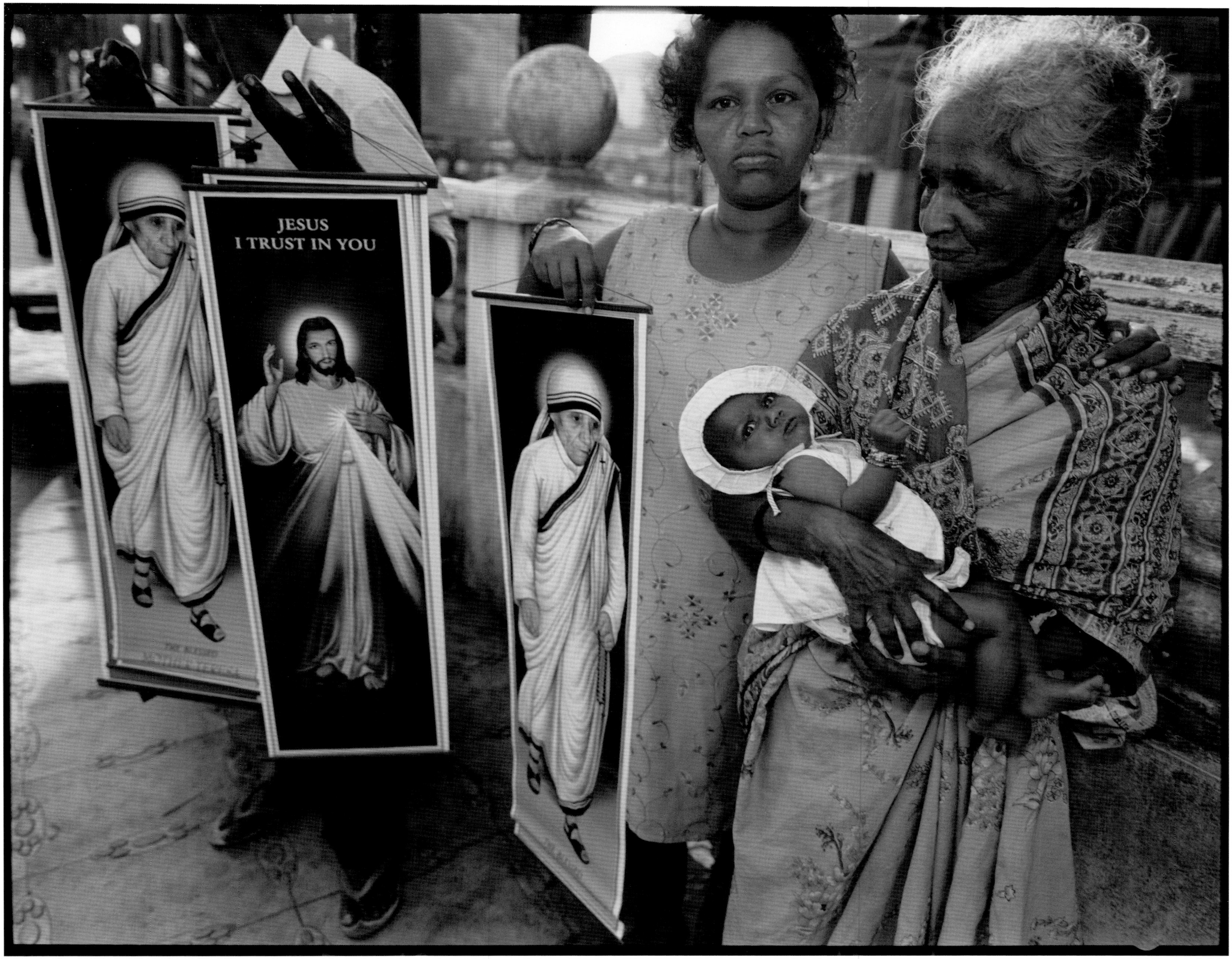

PLATE 57

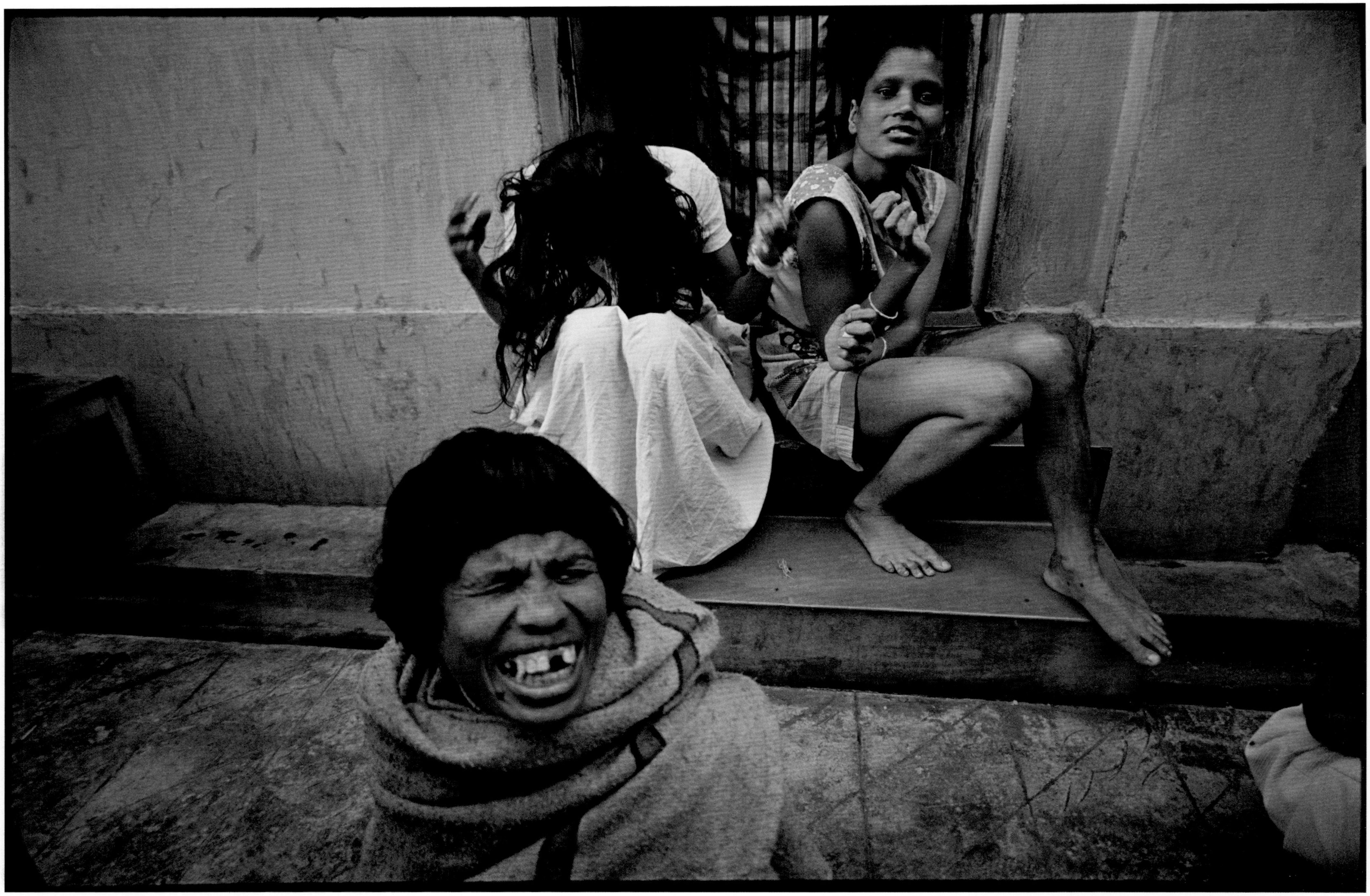

PLATE 58

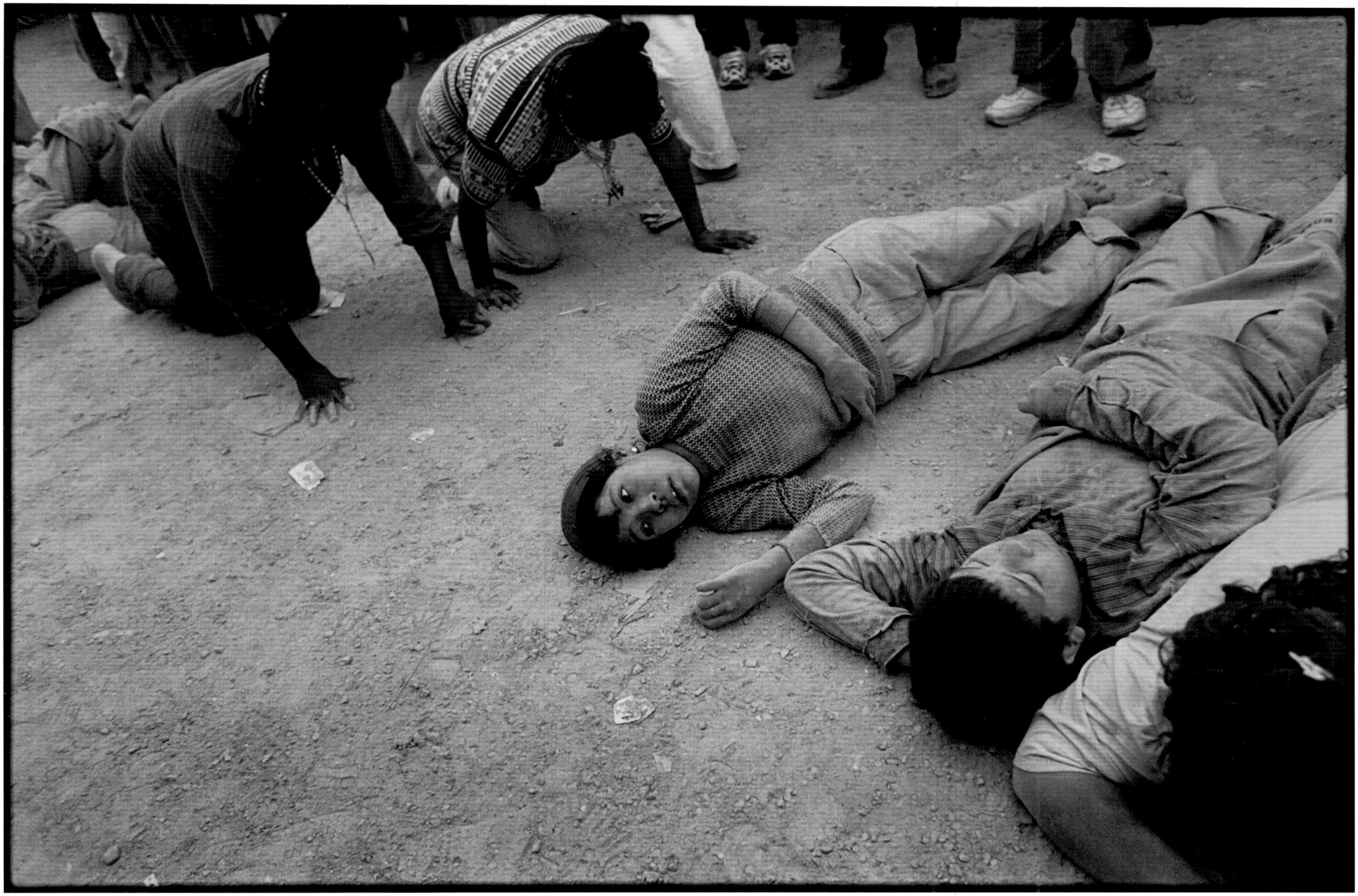

PLATE 59

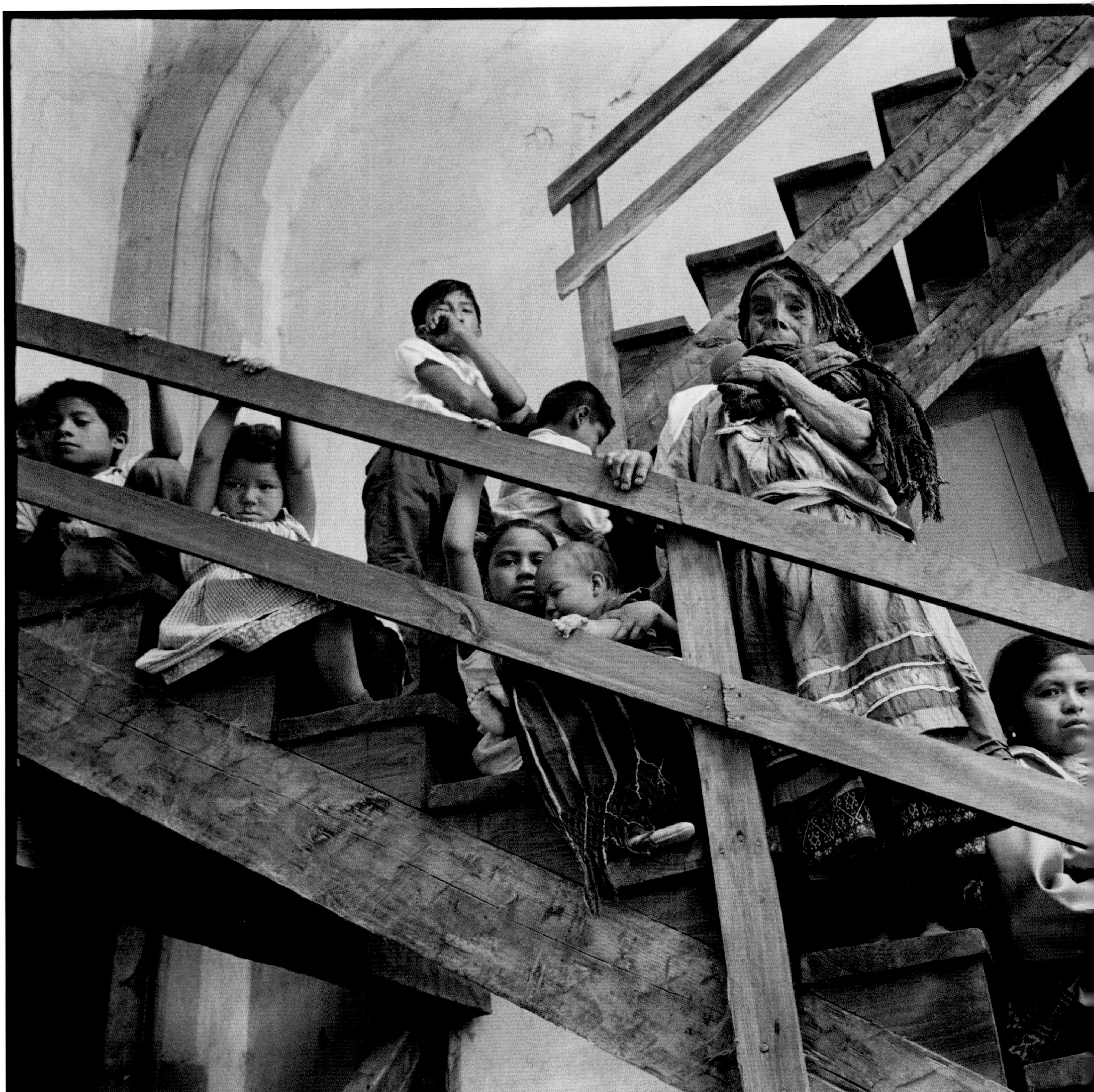

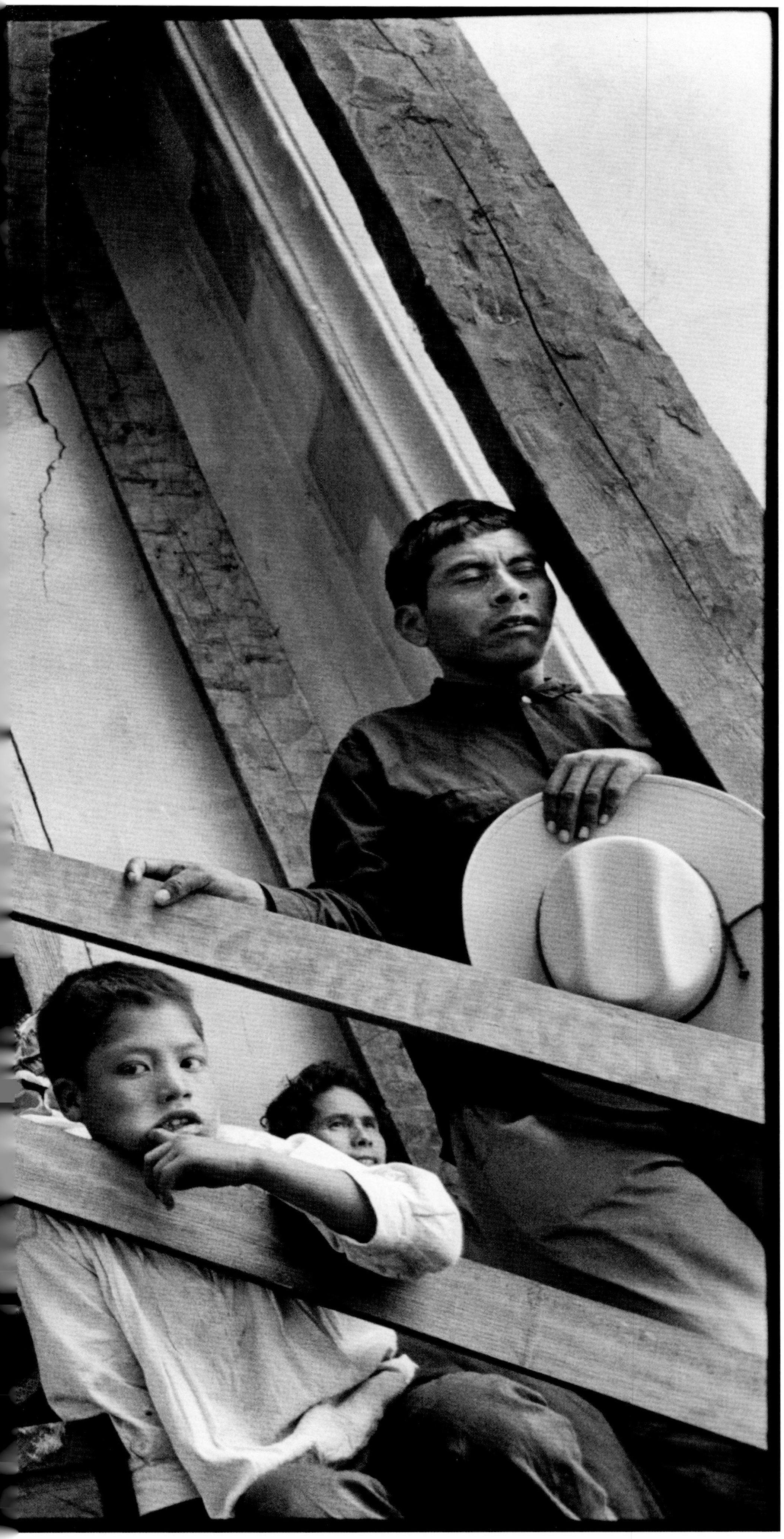

PLATE 60

PLATE 61

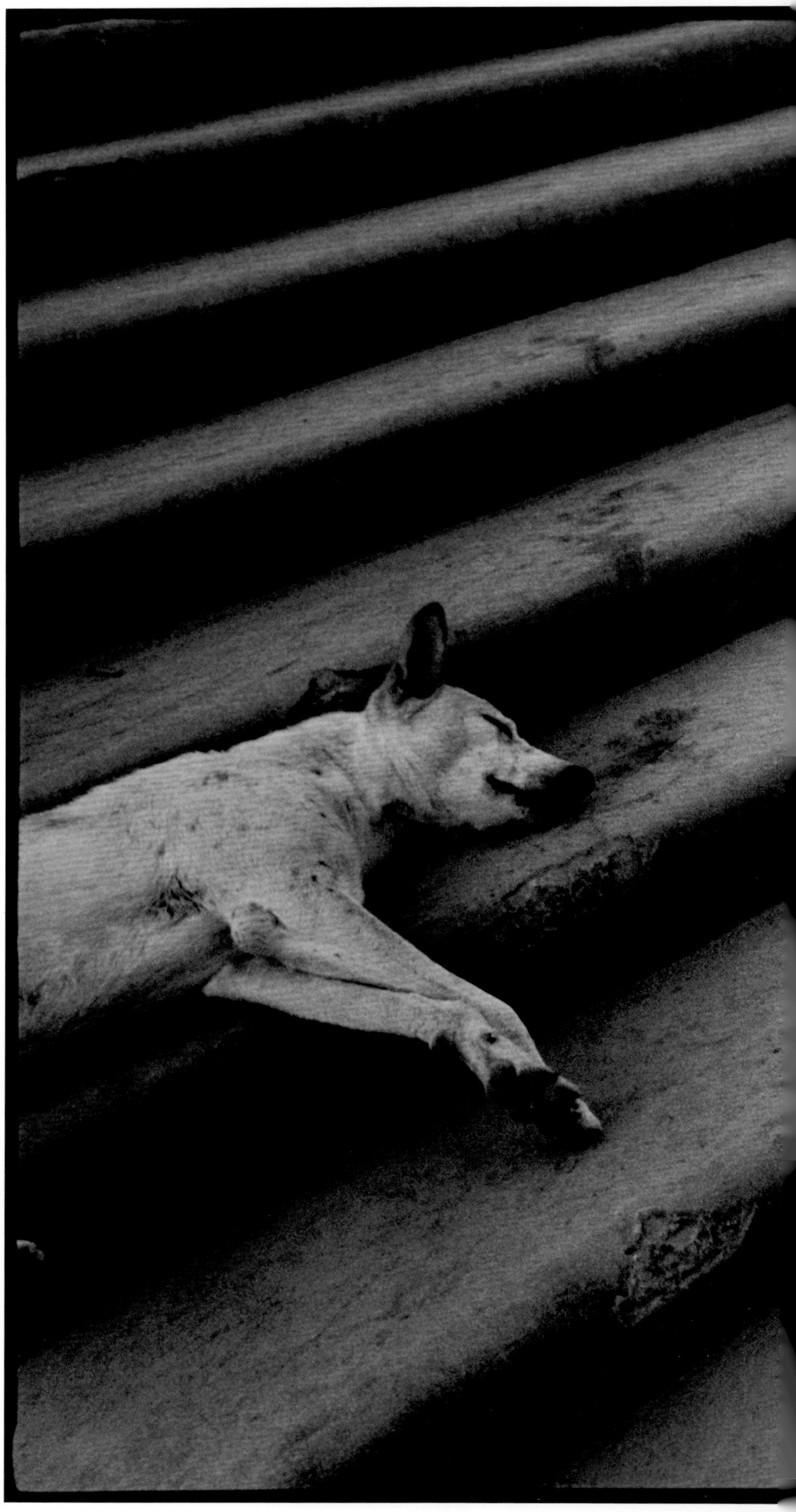

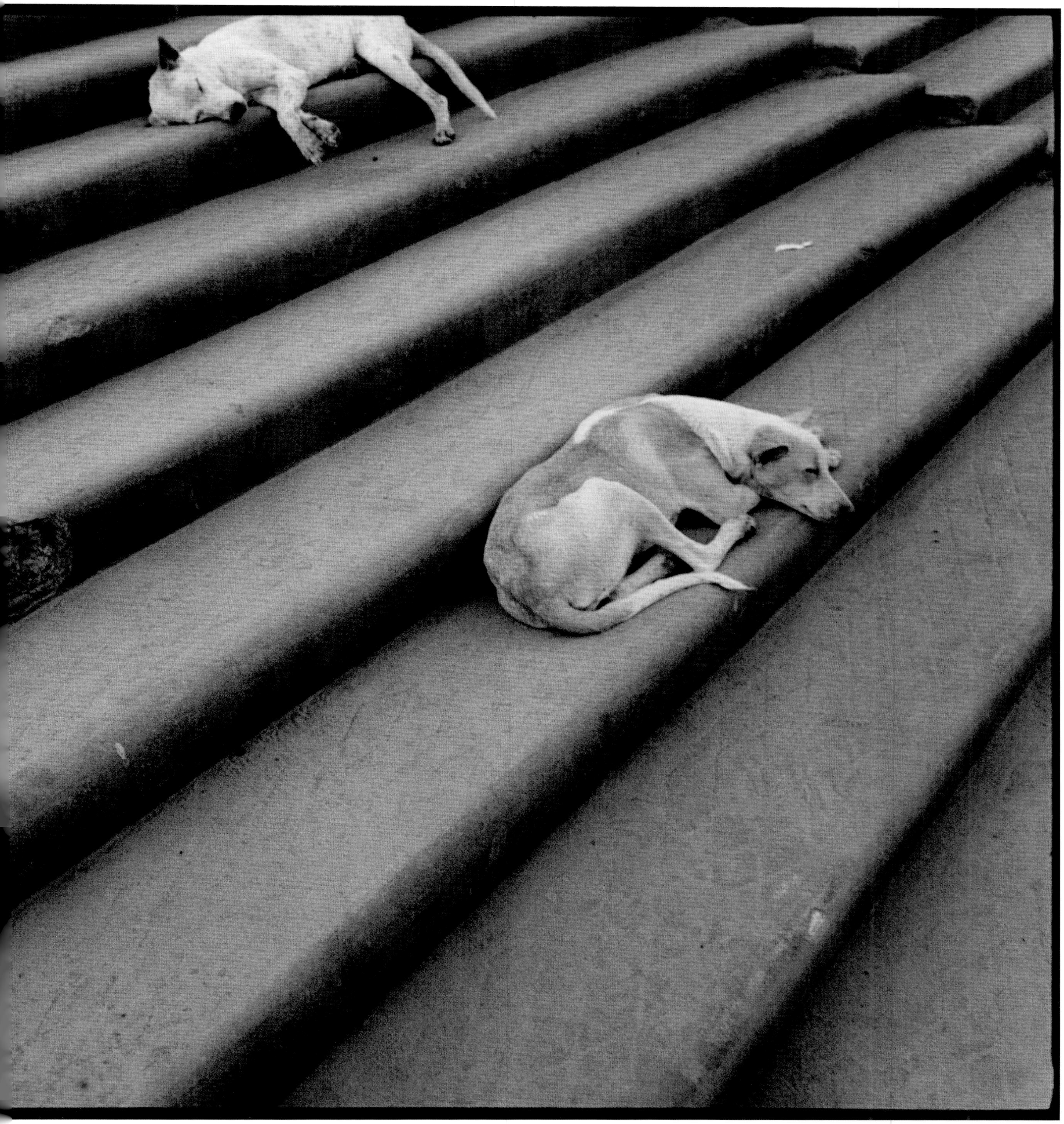

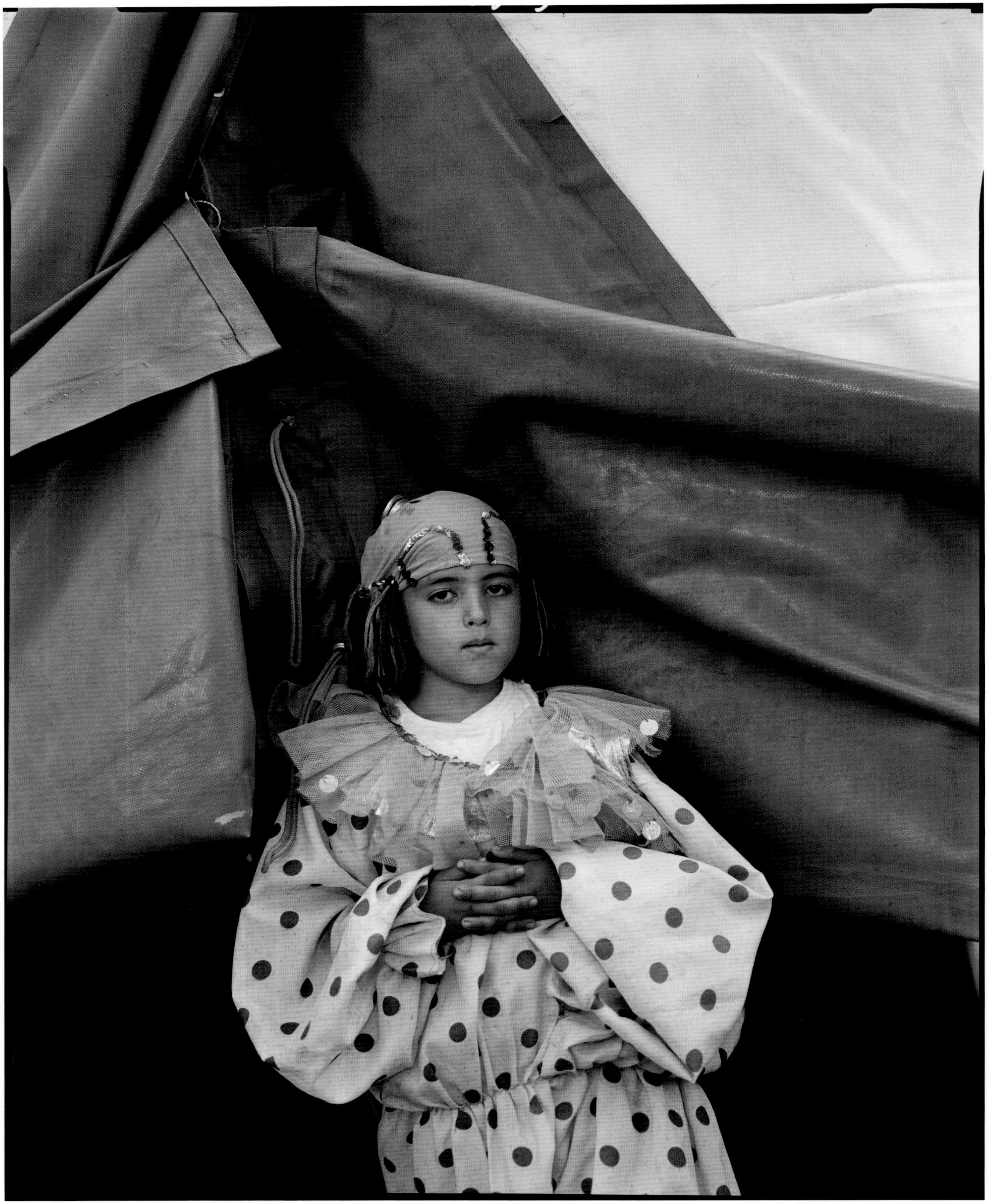

PLATE 62

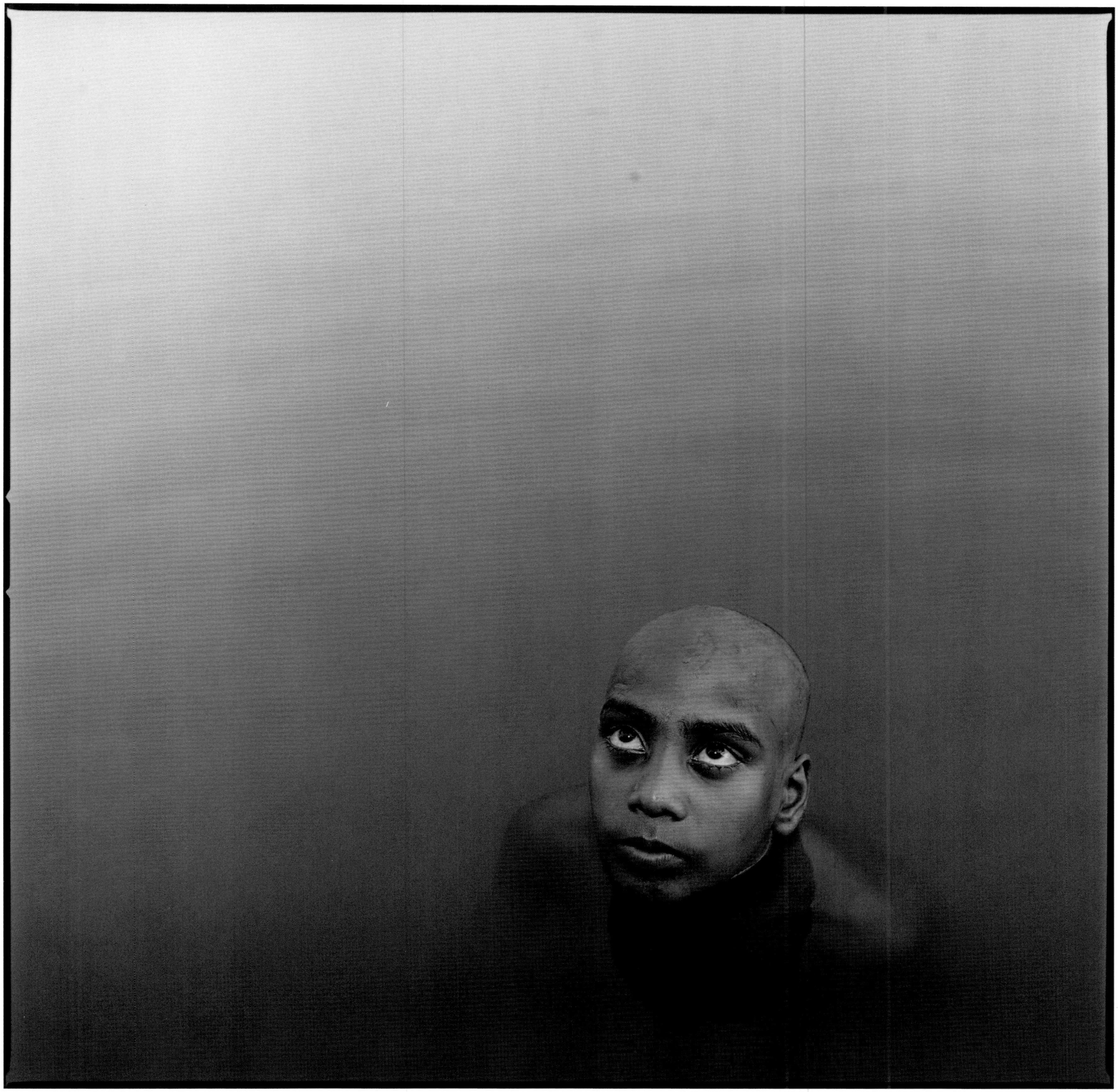

PLATE 63

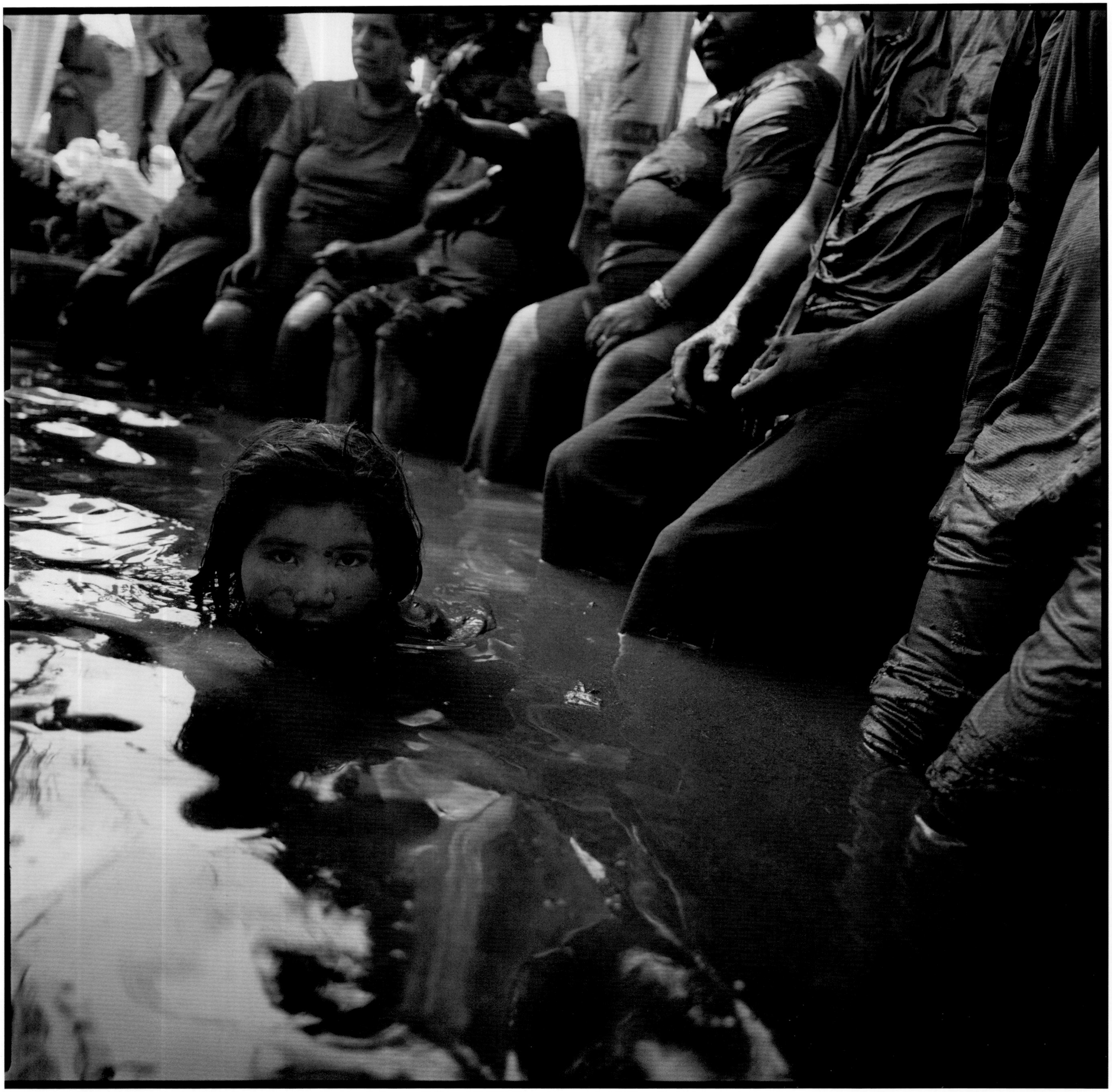

PLATE 64

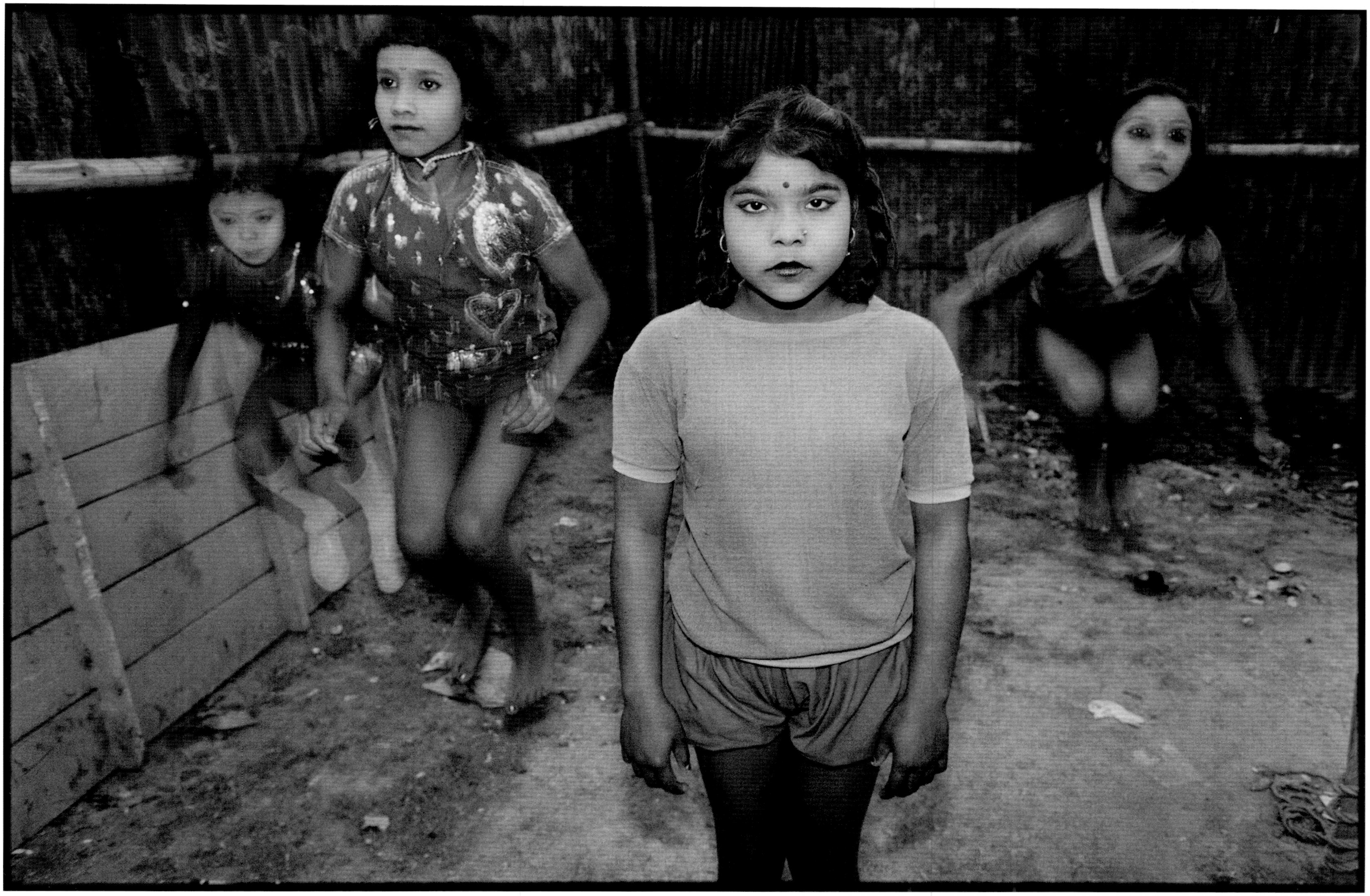

PLATE 65

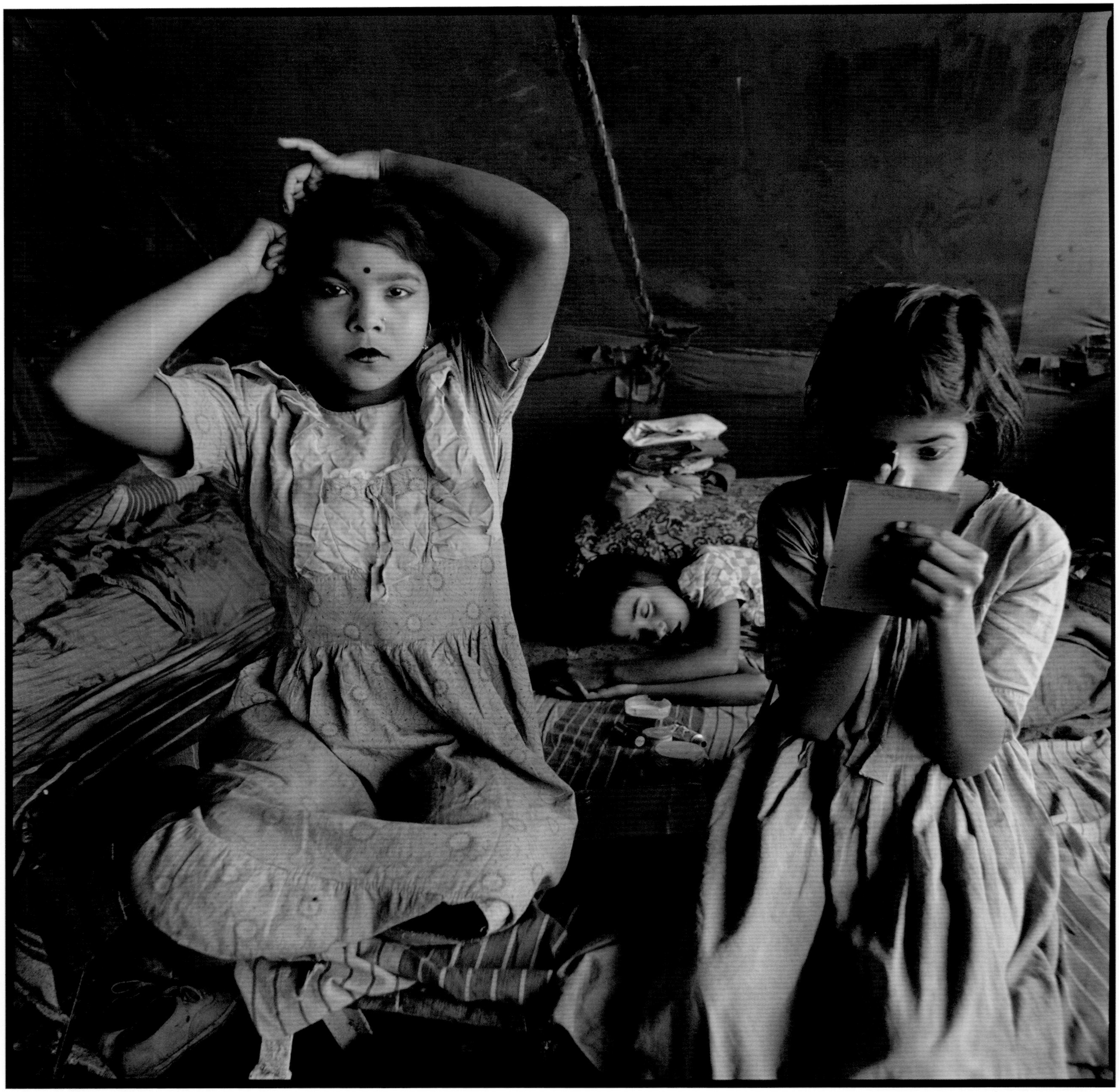

PLATE 66

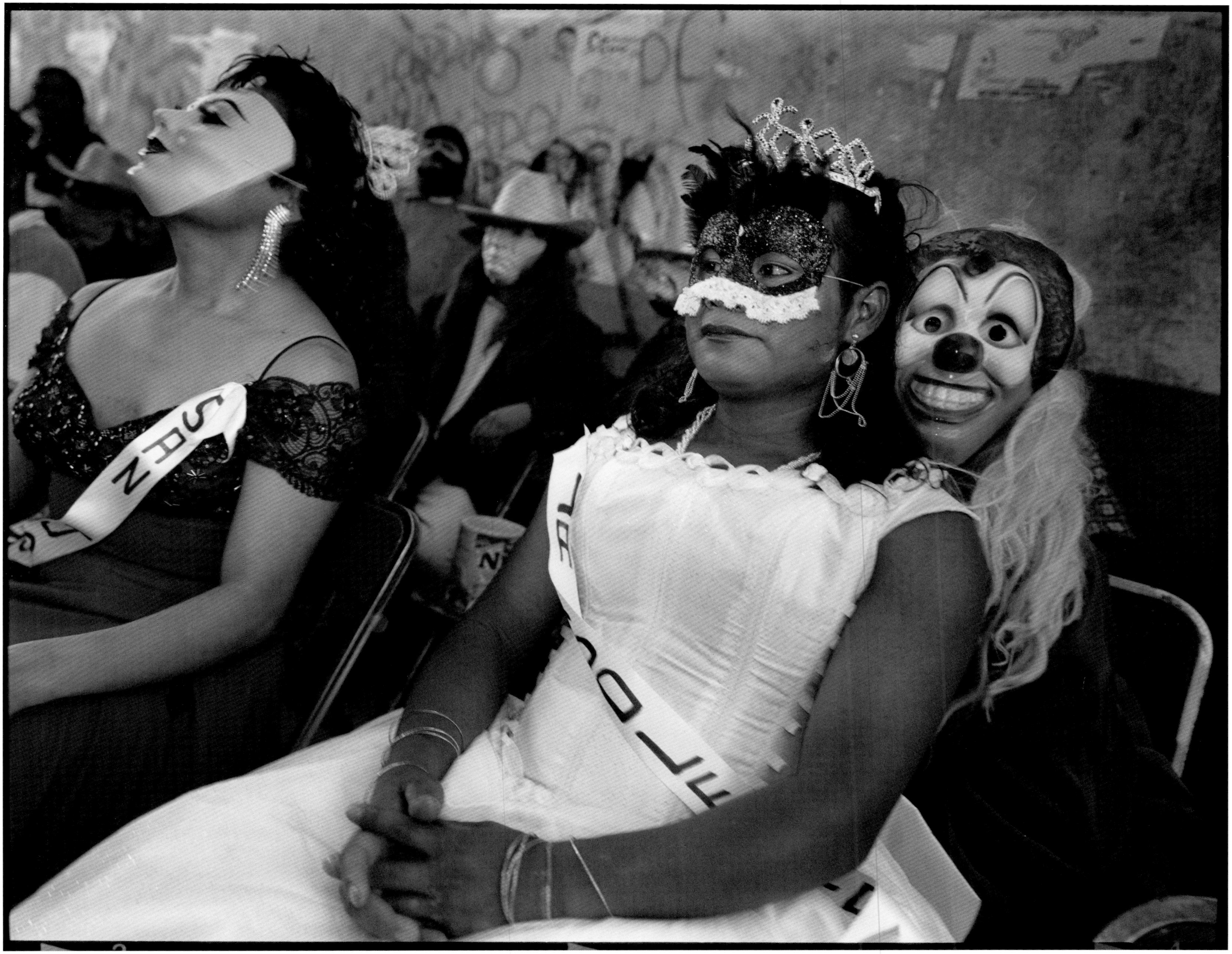

PLATE 67

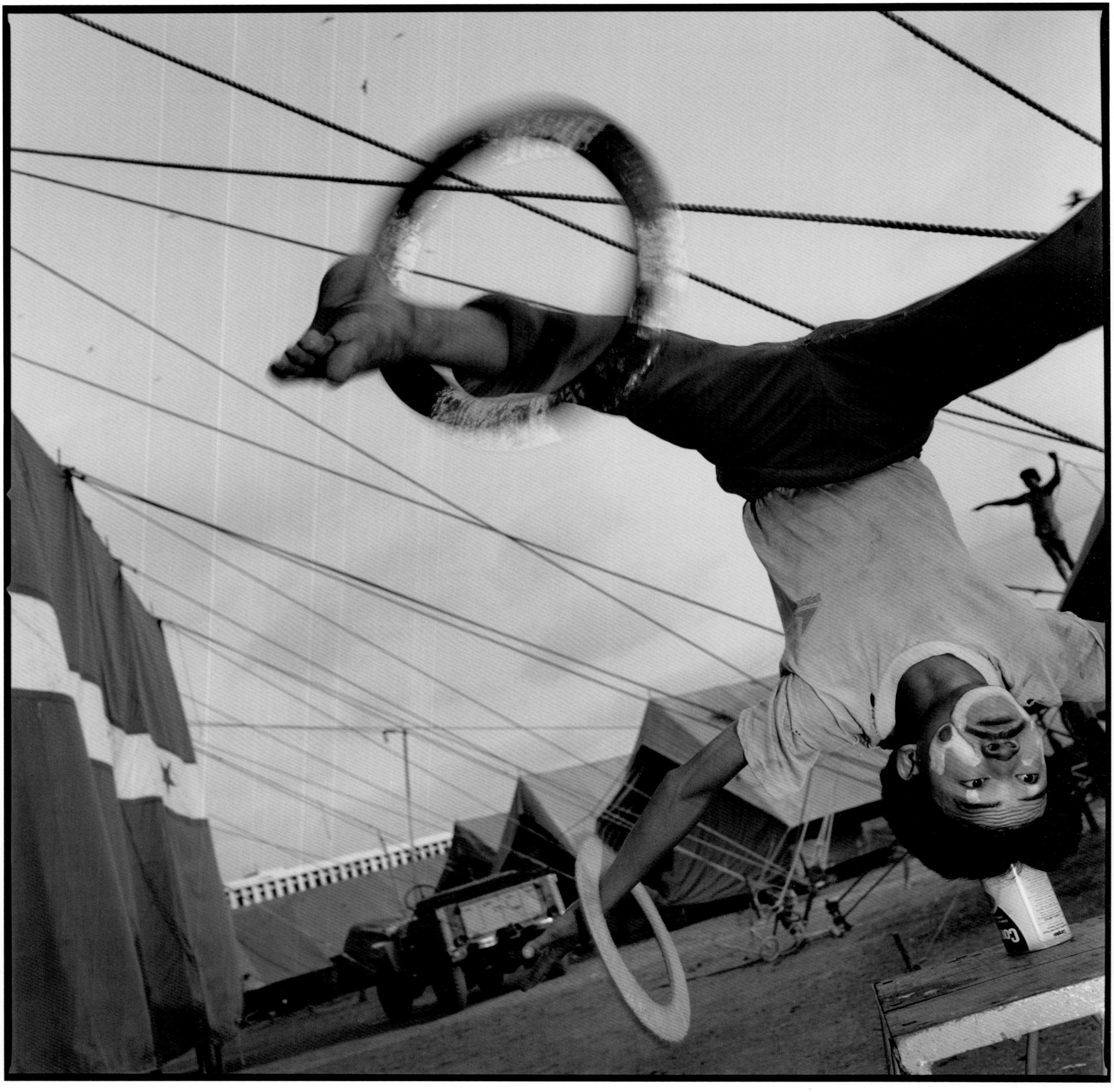

PLATE 68

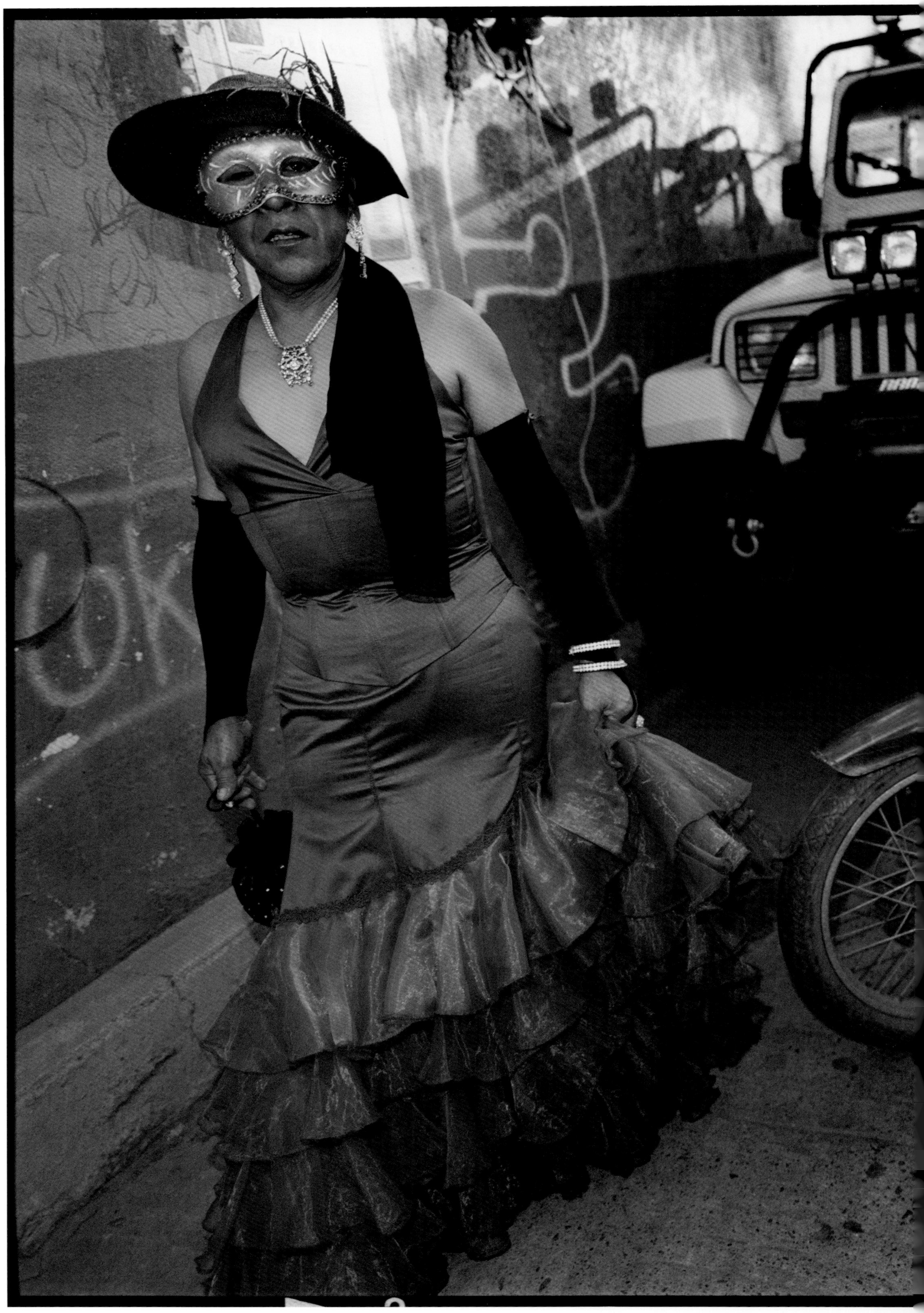

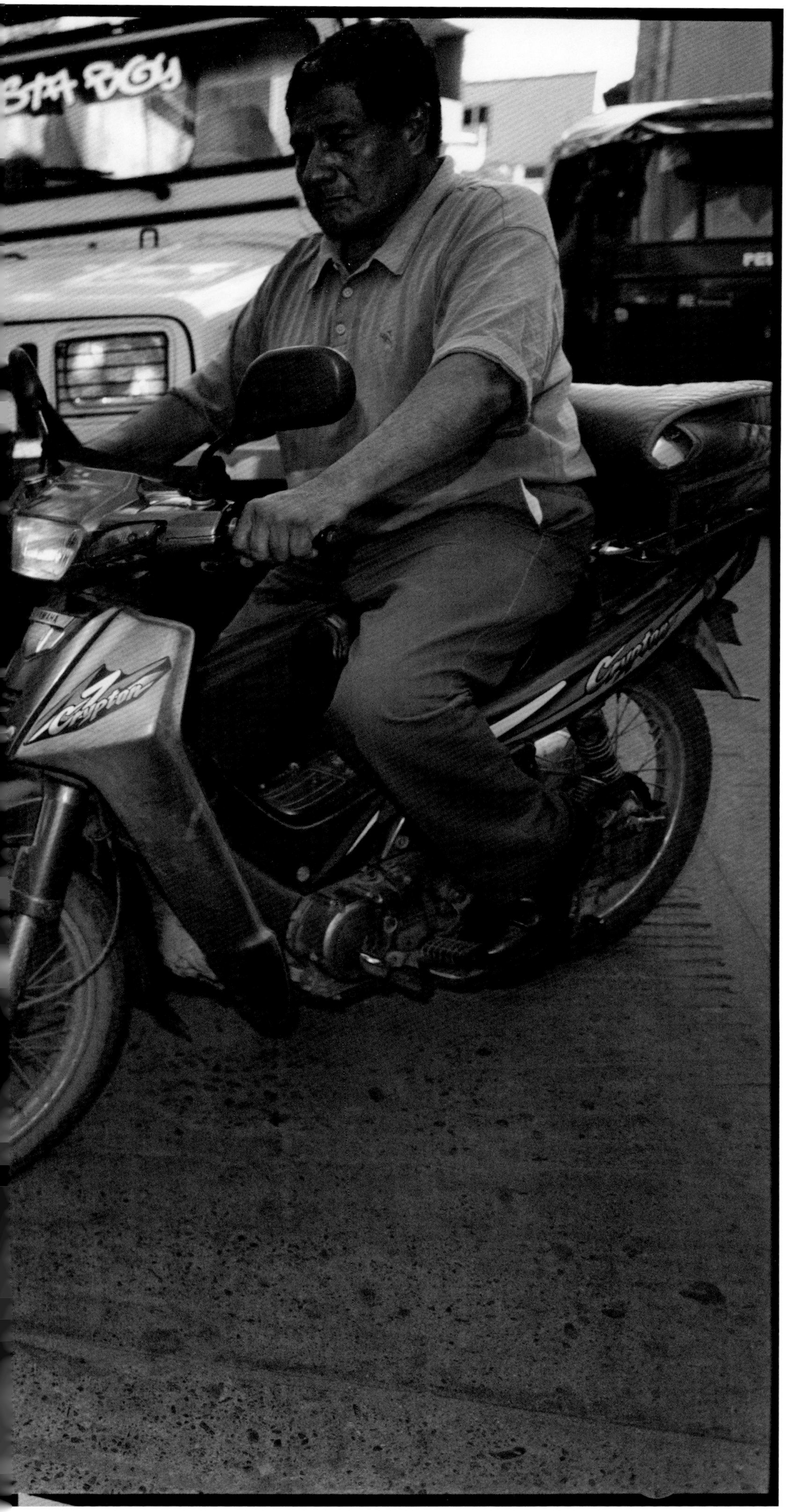

PLATE 69

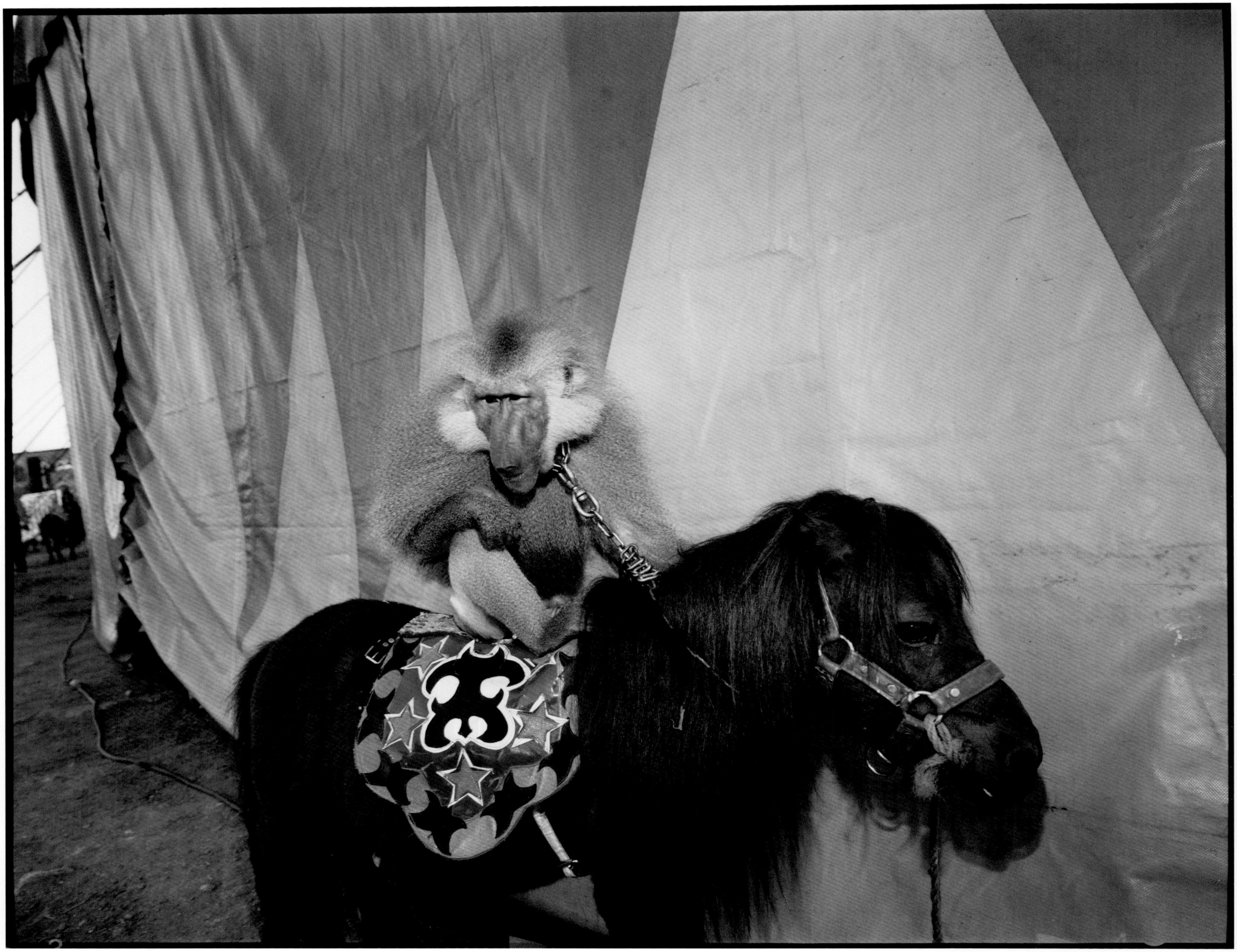

PLATE 70

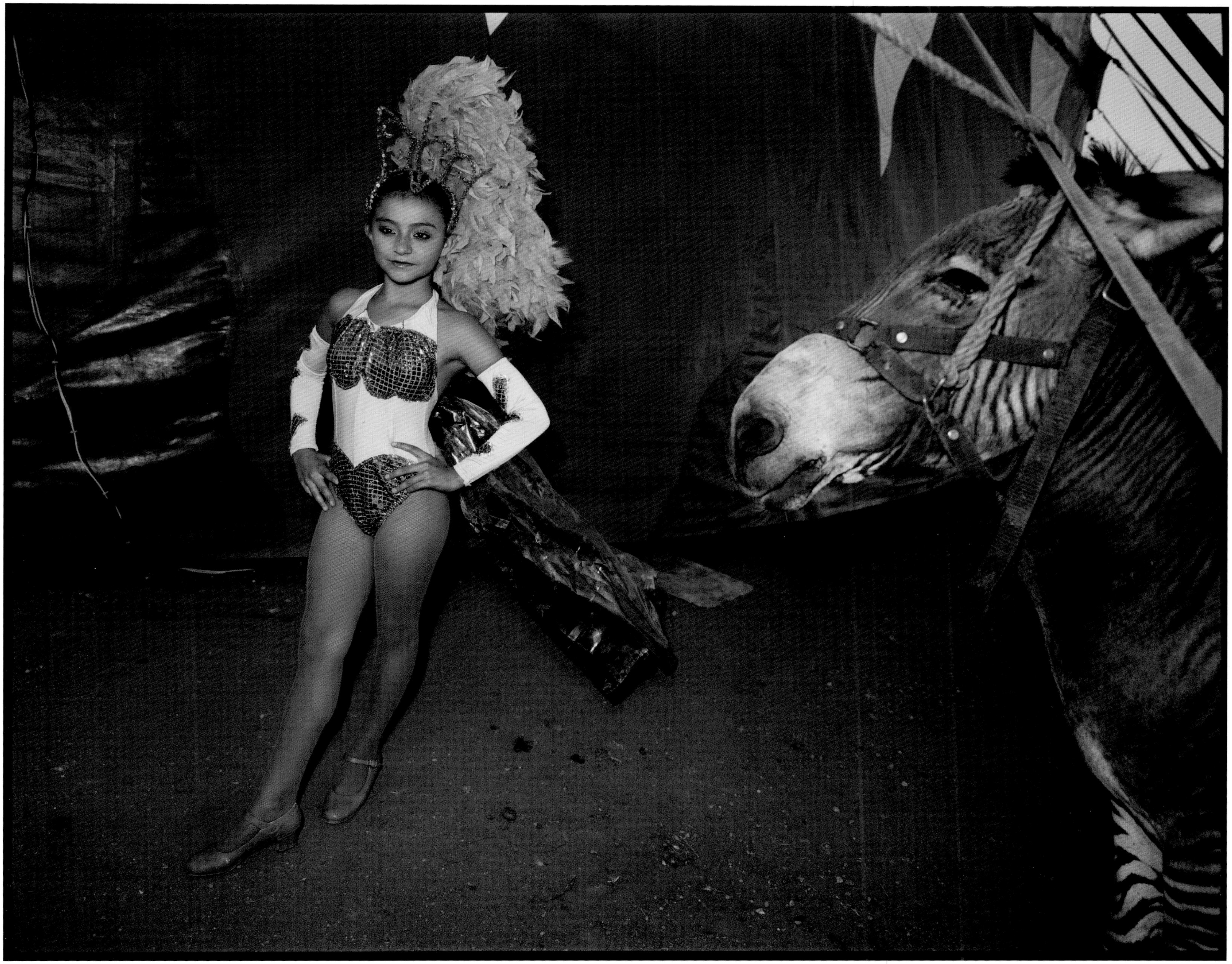

PLATE 71

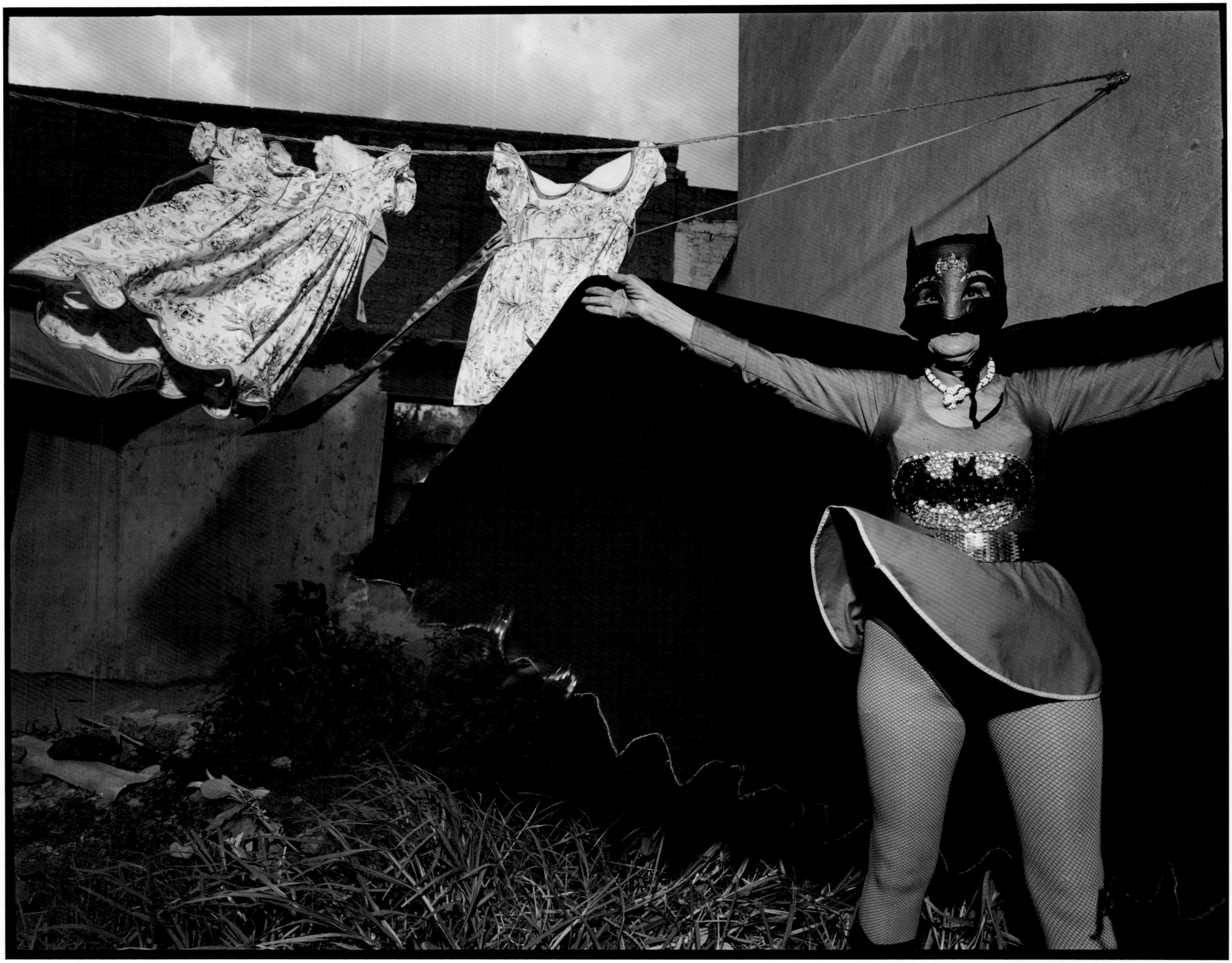

PLATE 72

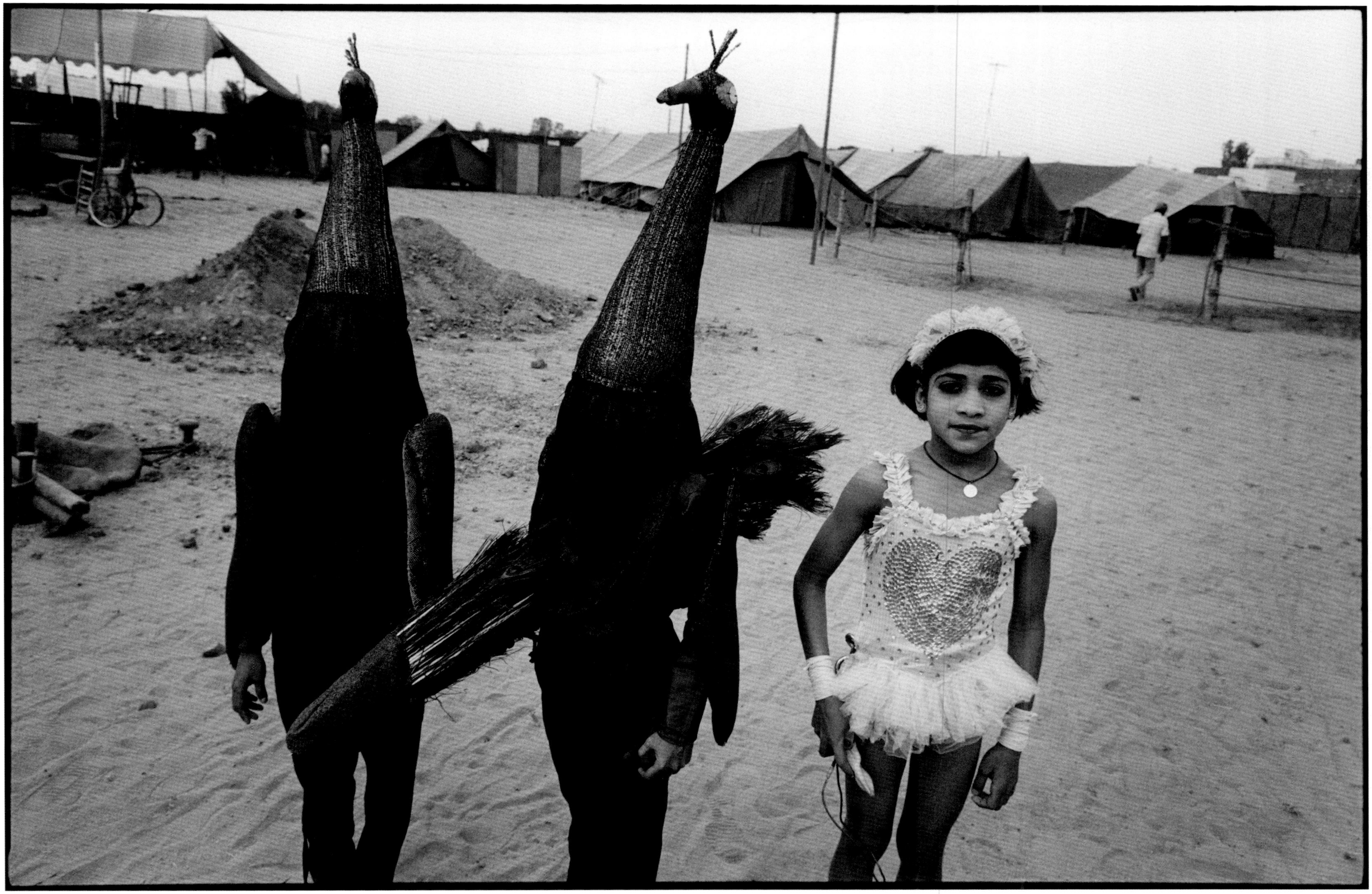

PLATE 73

PLATE 74

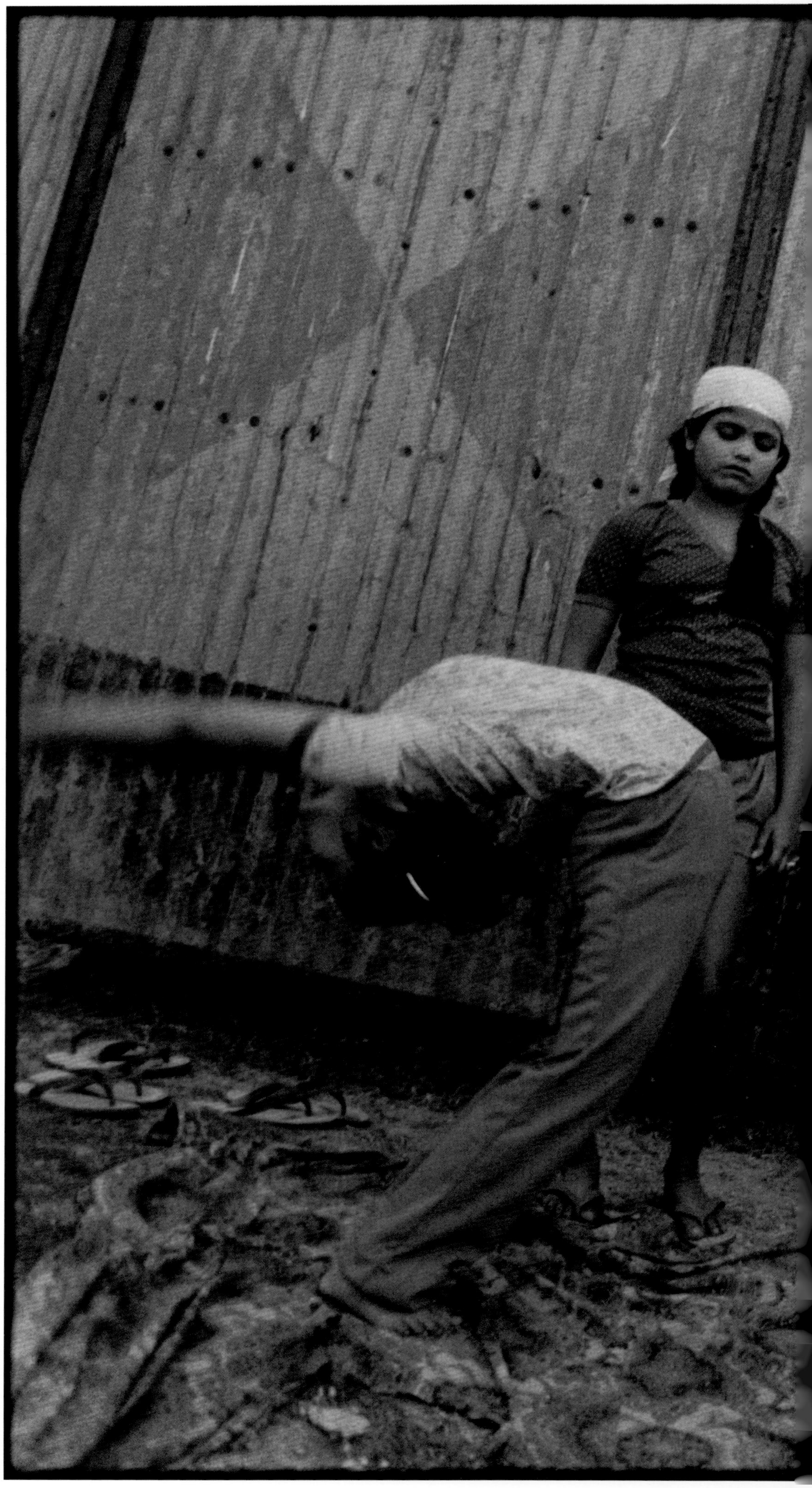

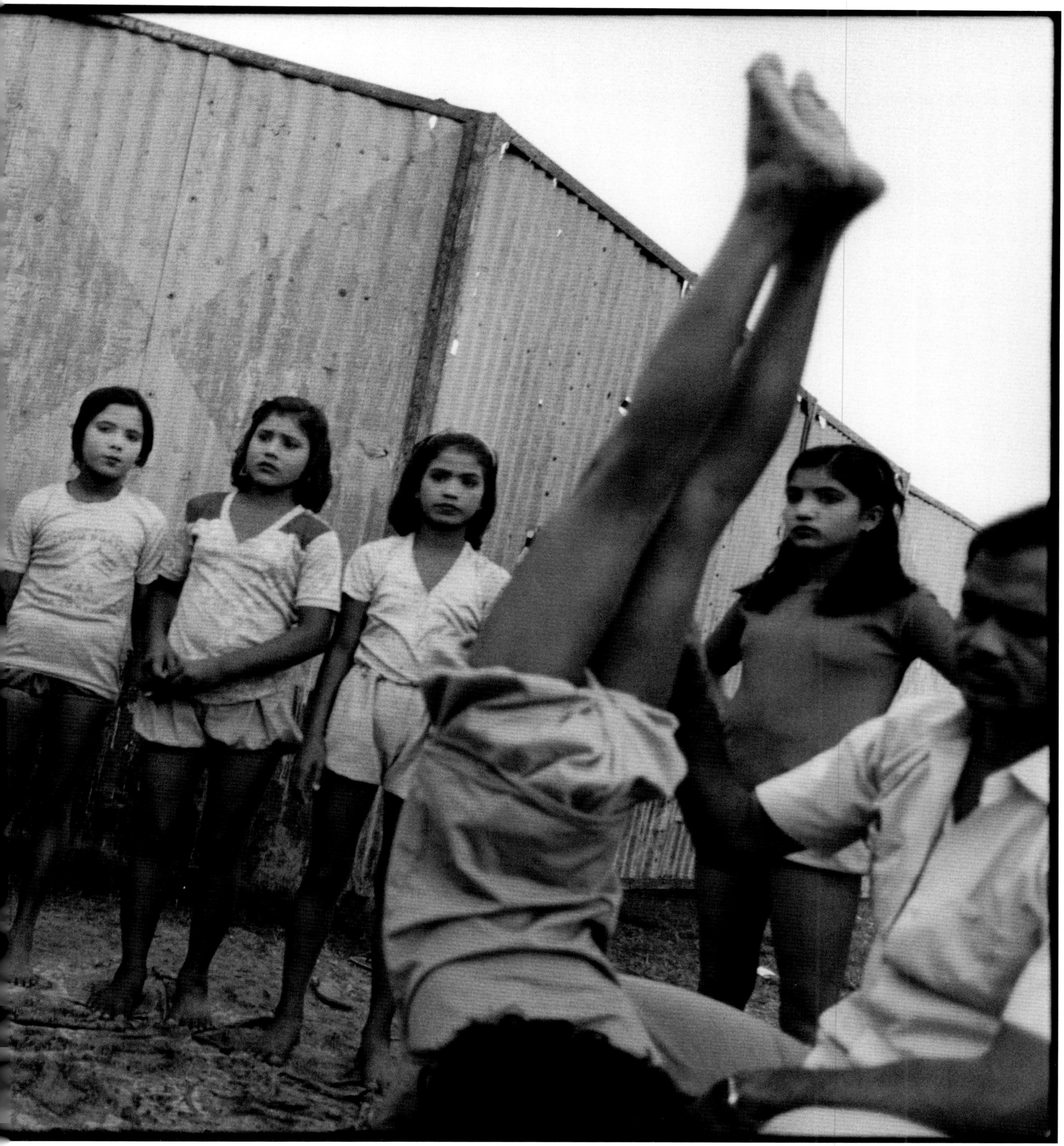

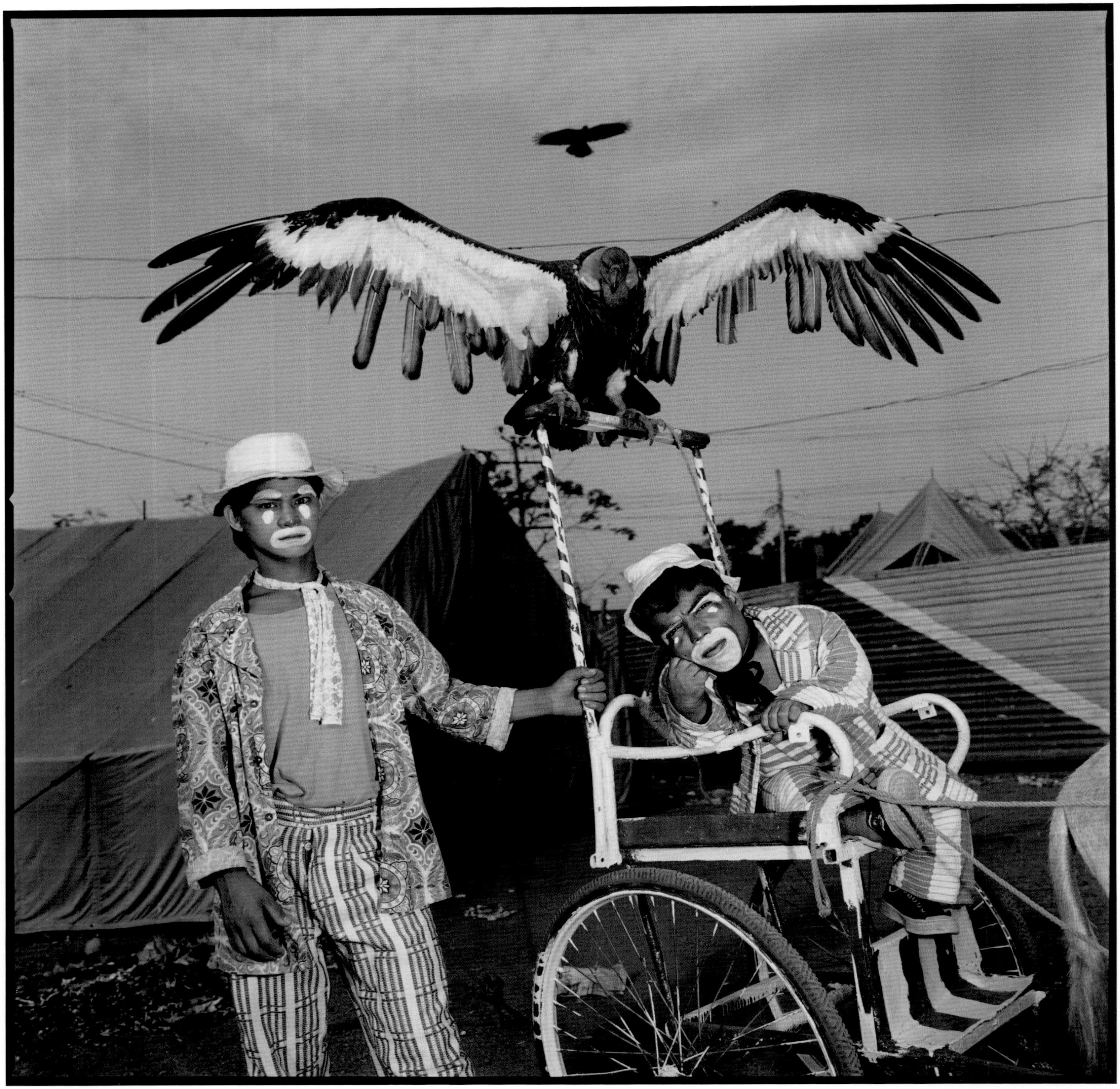

PLATE 75

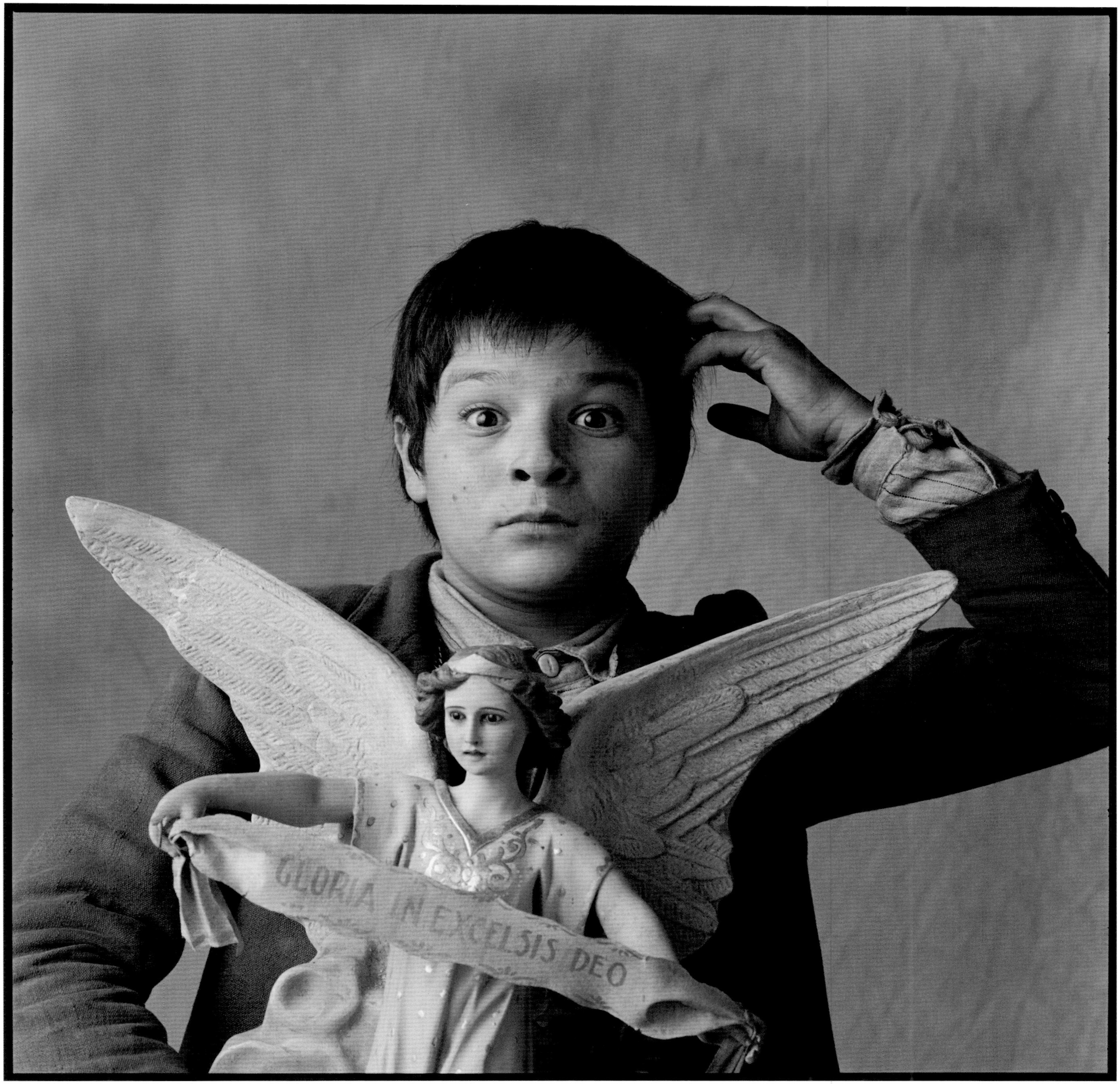

PLATE 76

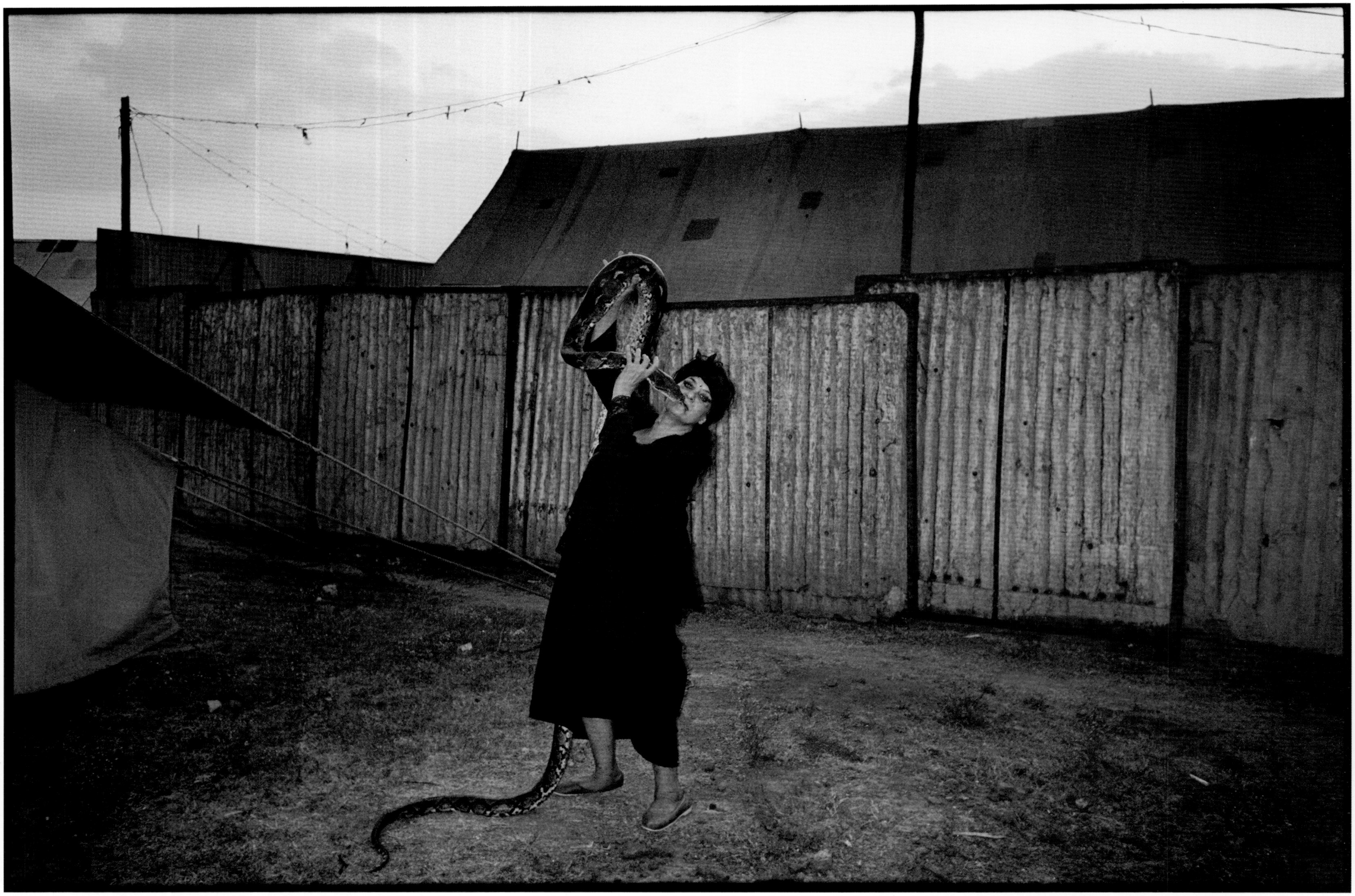

PLATE 77

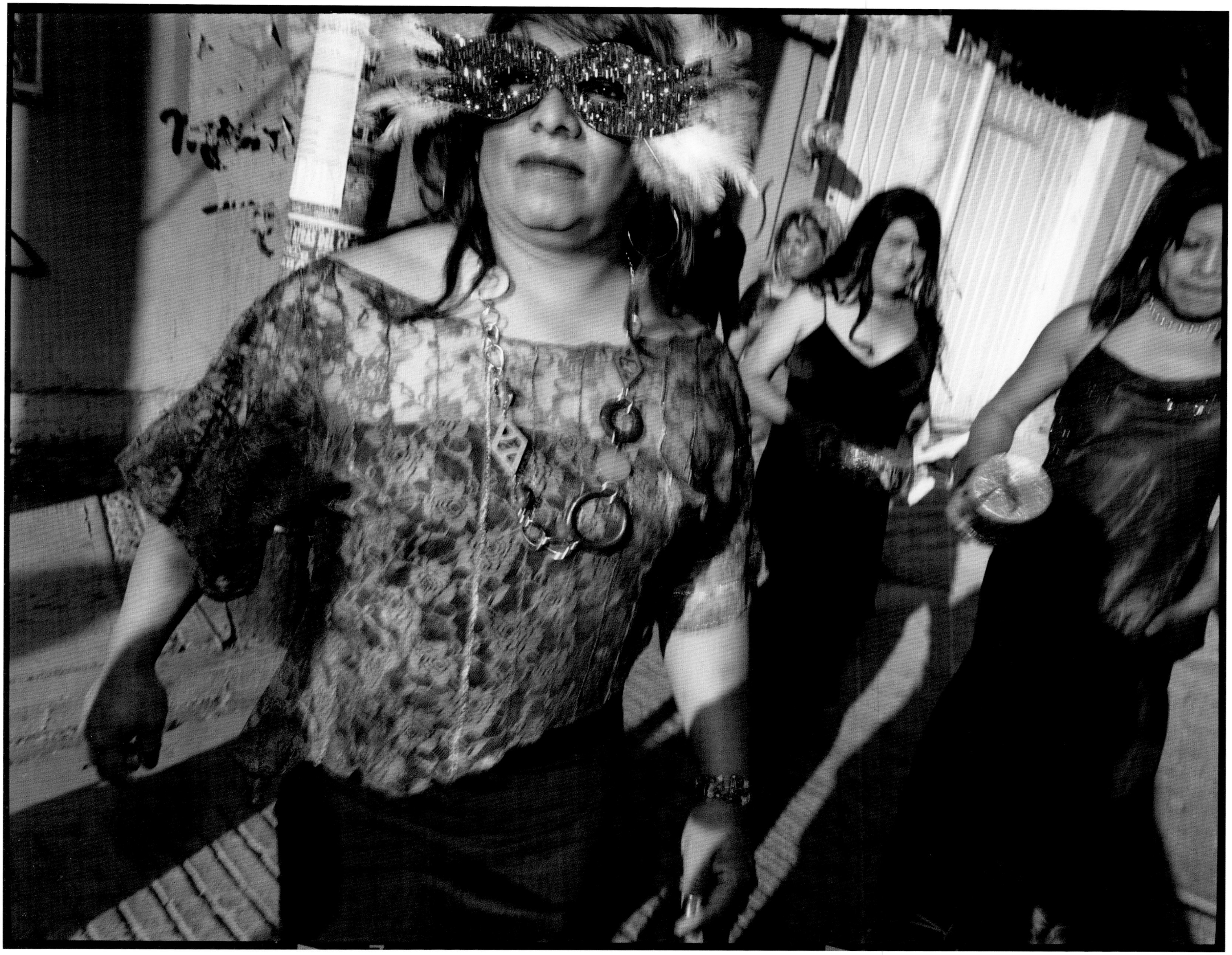

PLATE 78

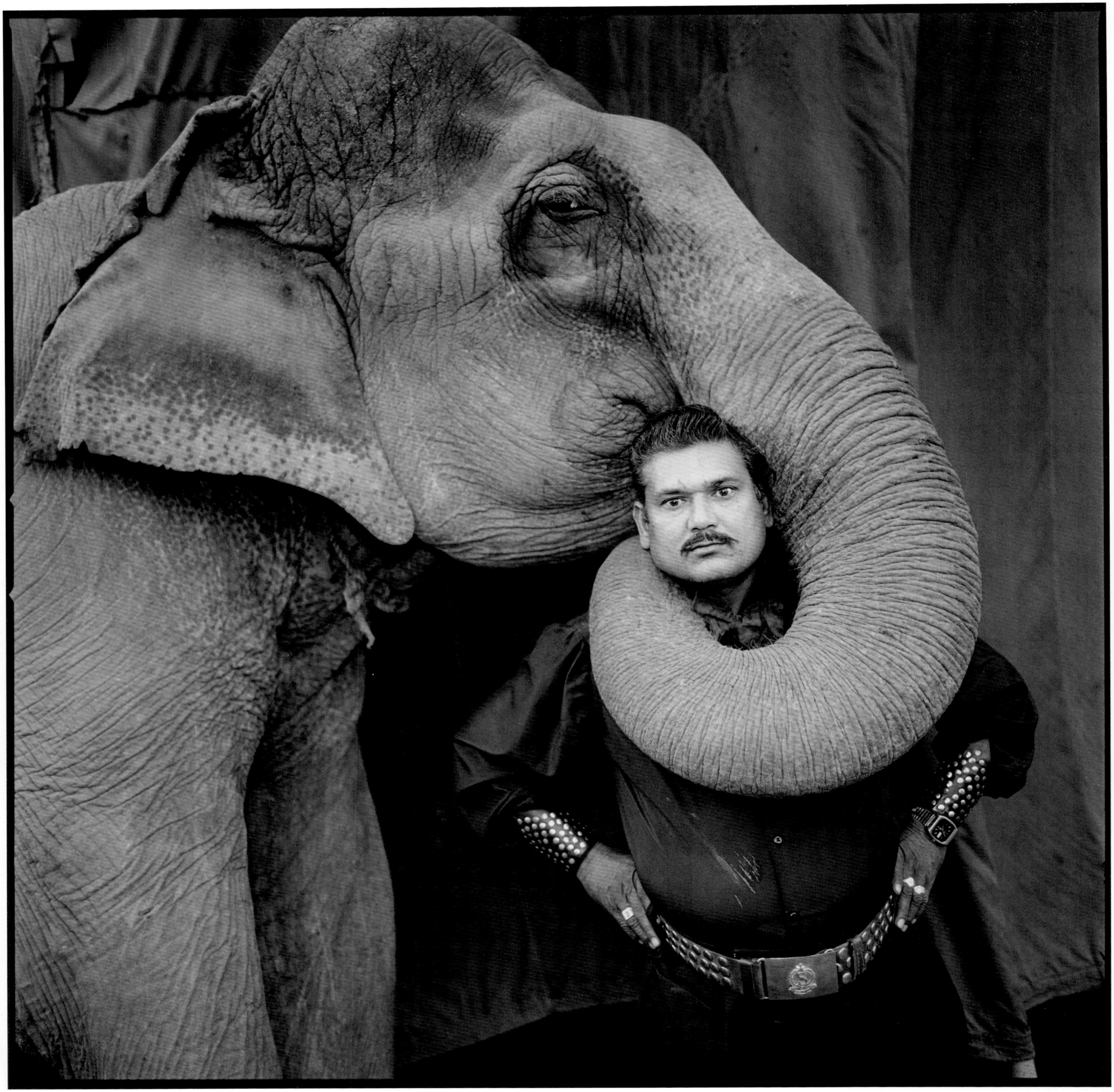

PLATE 79

PLATE 80

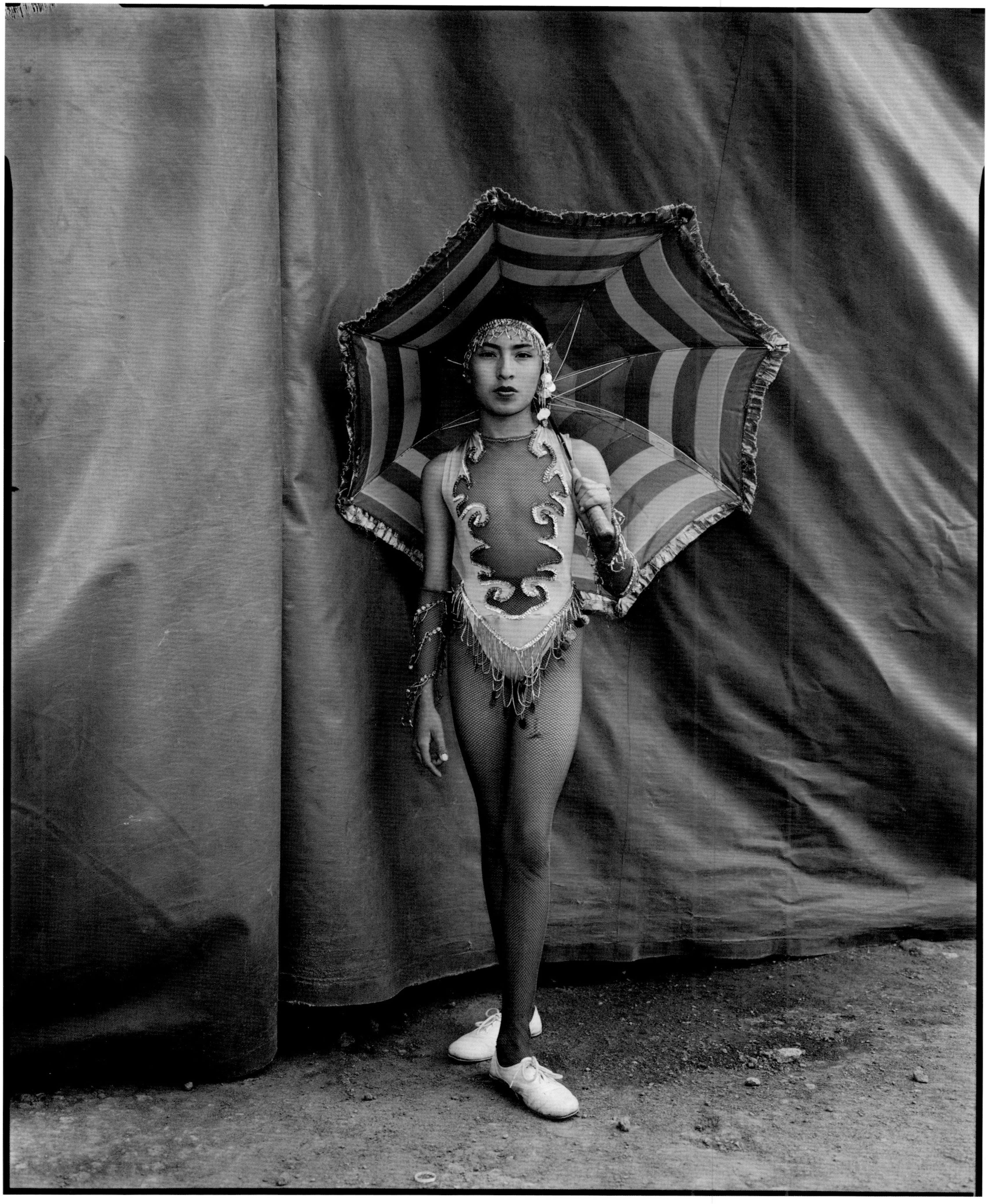

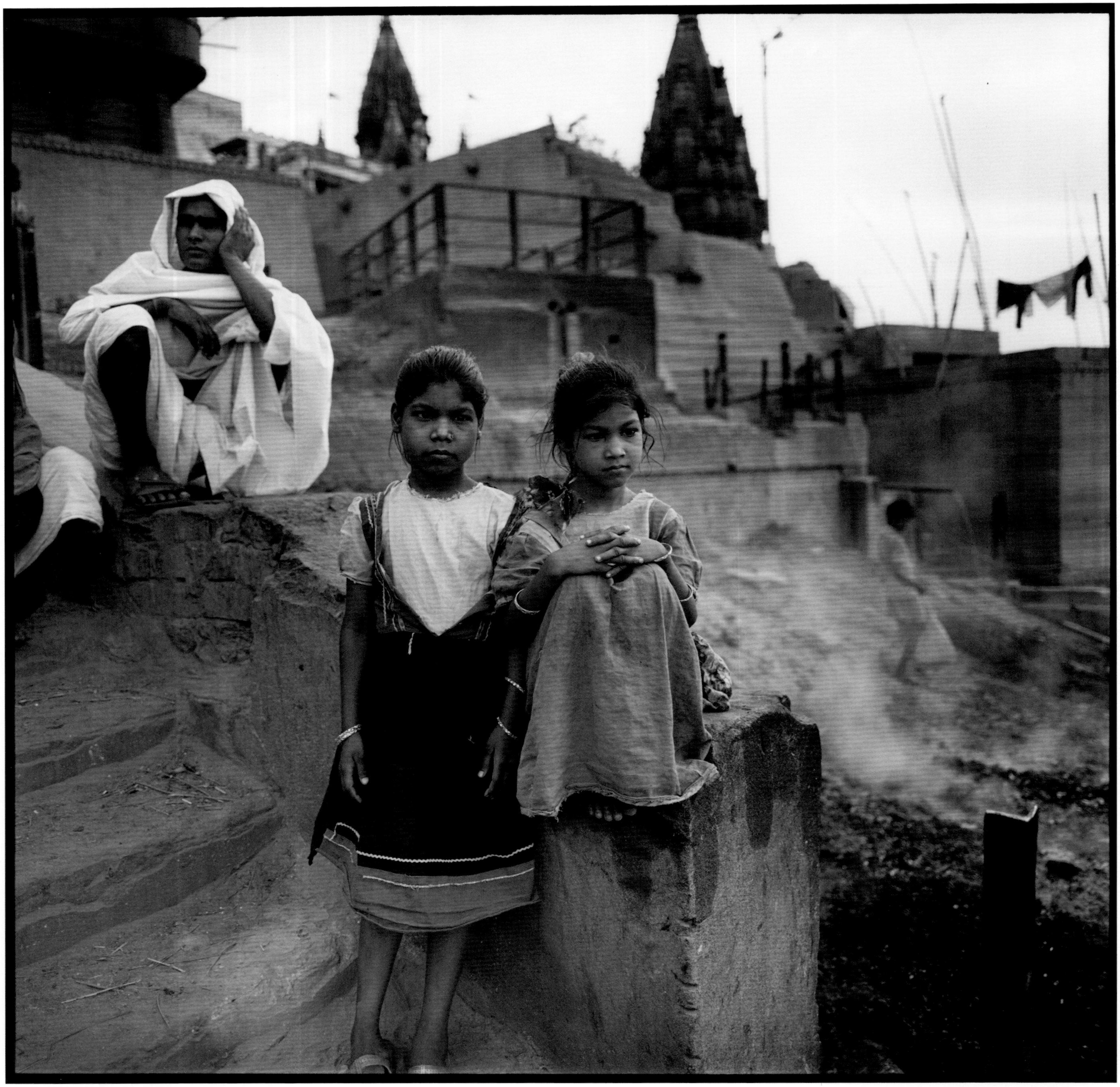

PLATE 81

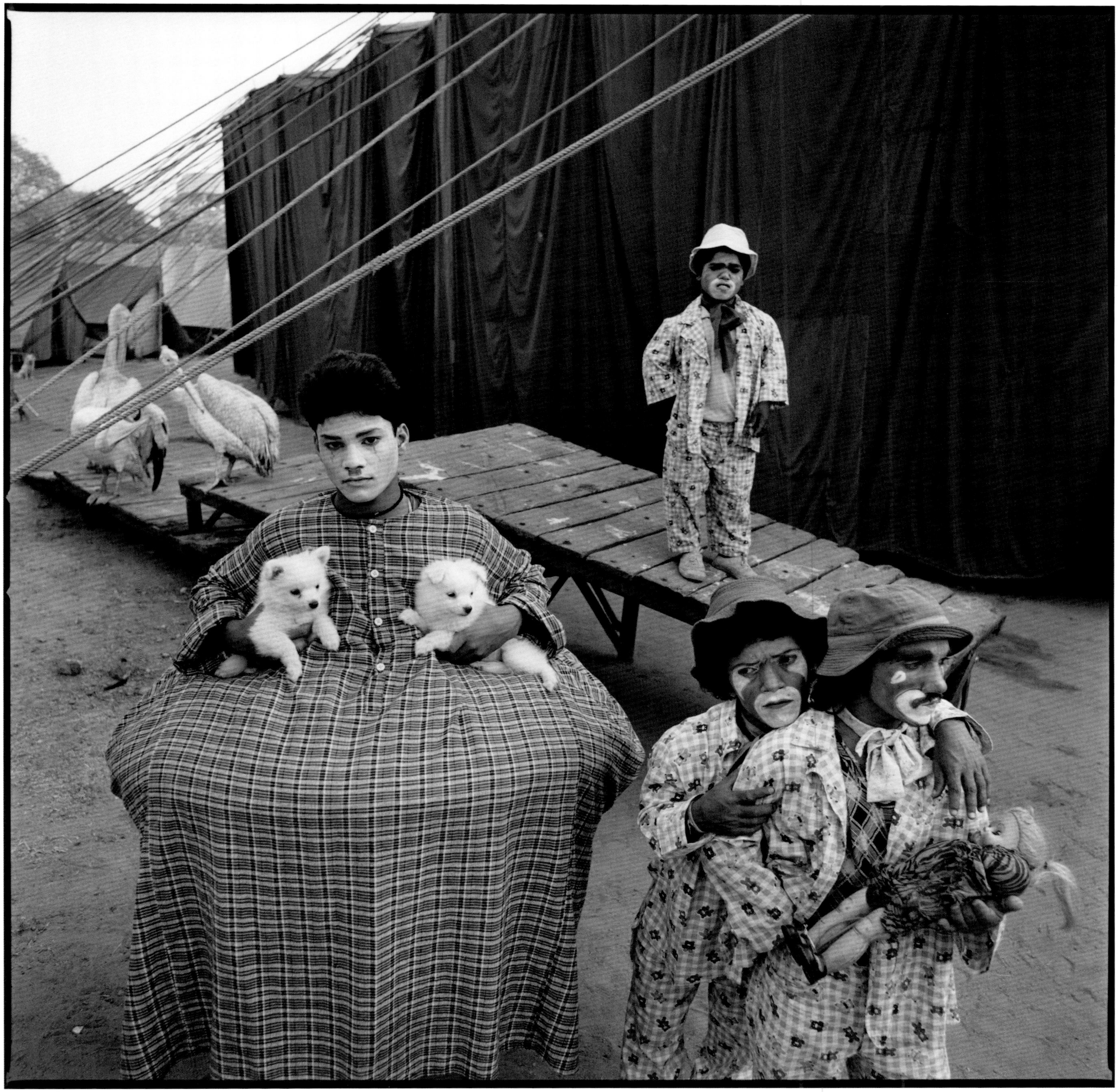

PLATE 82

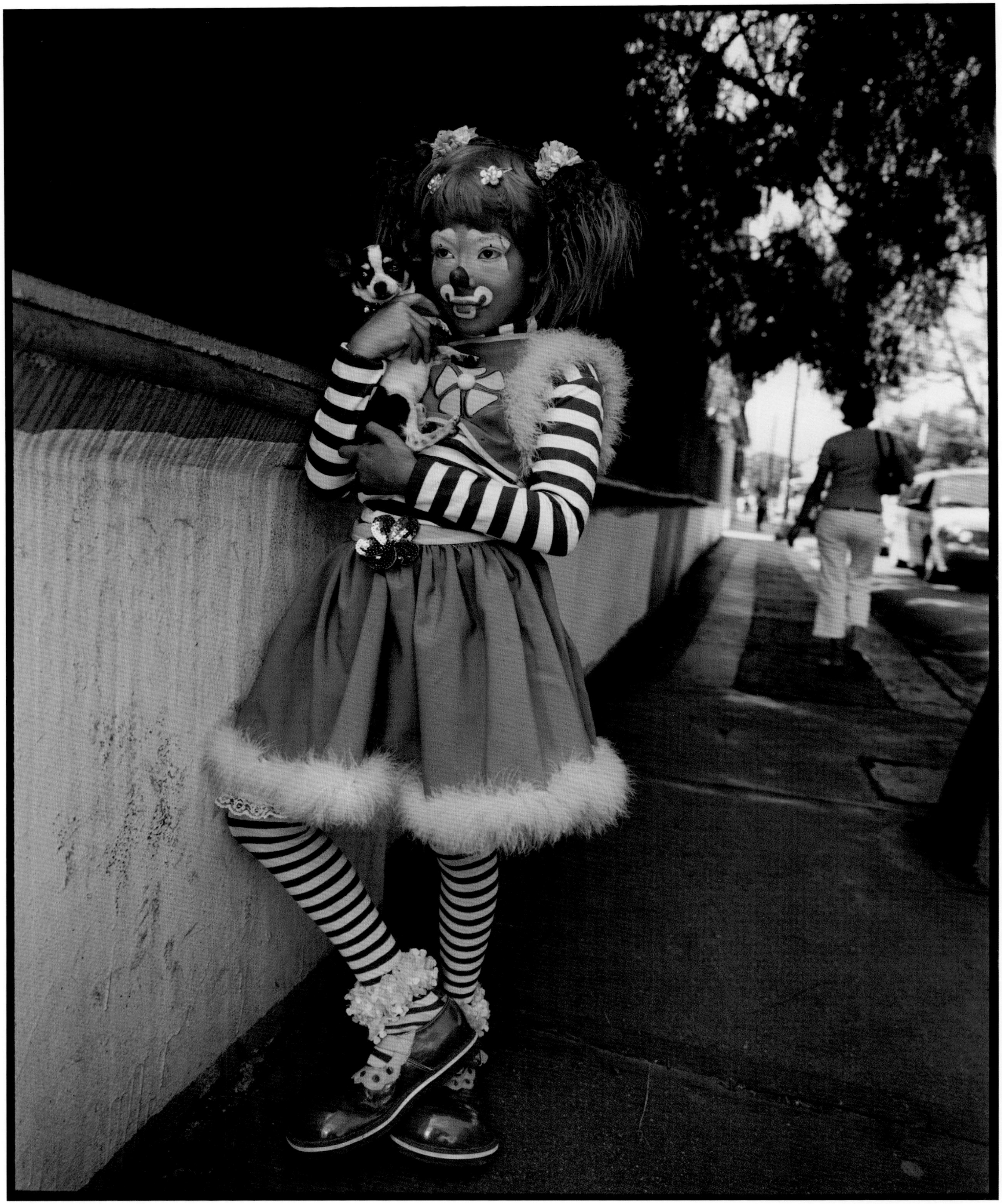

PLATE 83

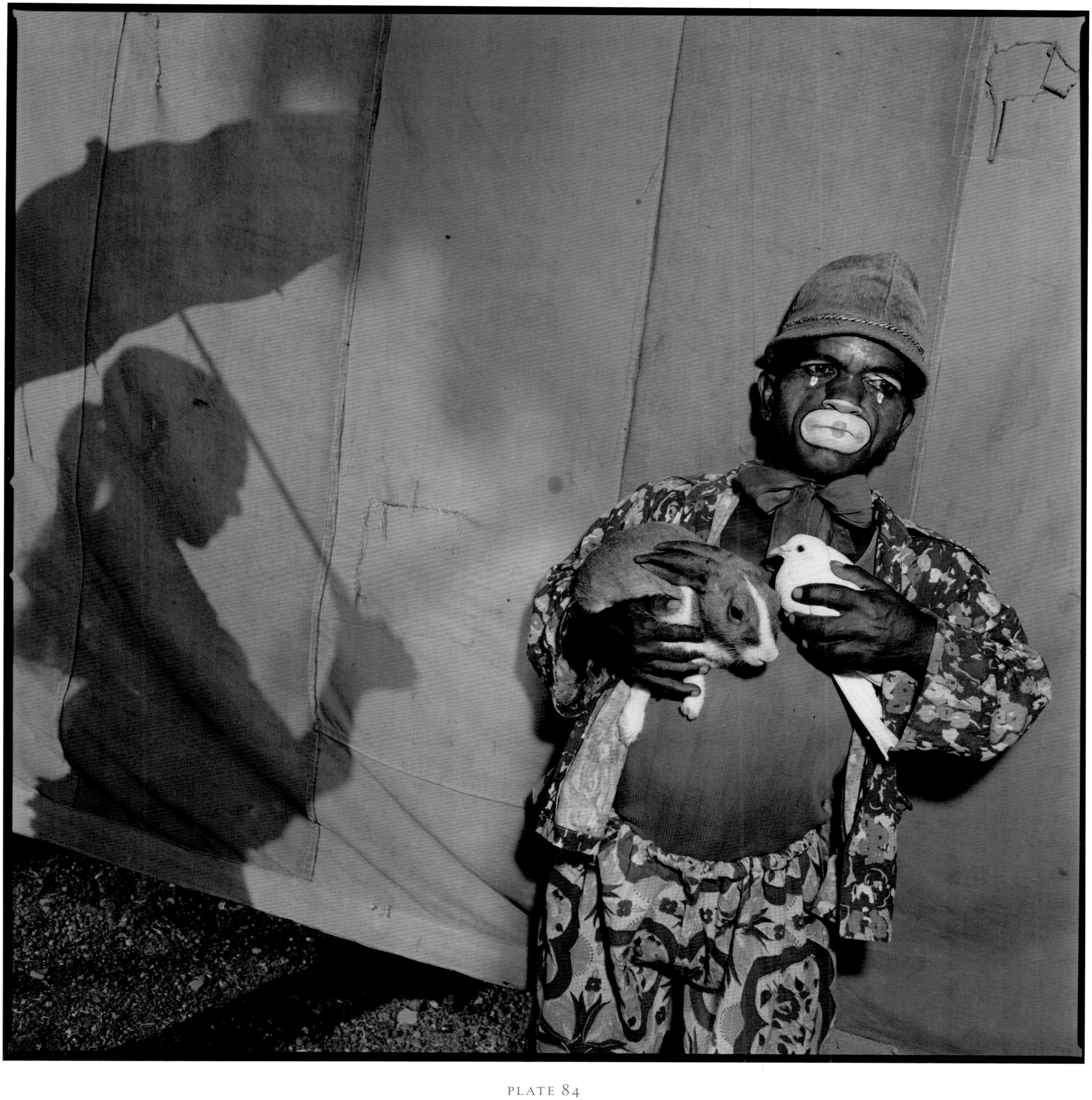

PLATE 84

PLATE 85

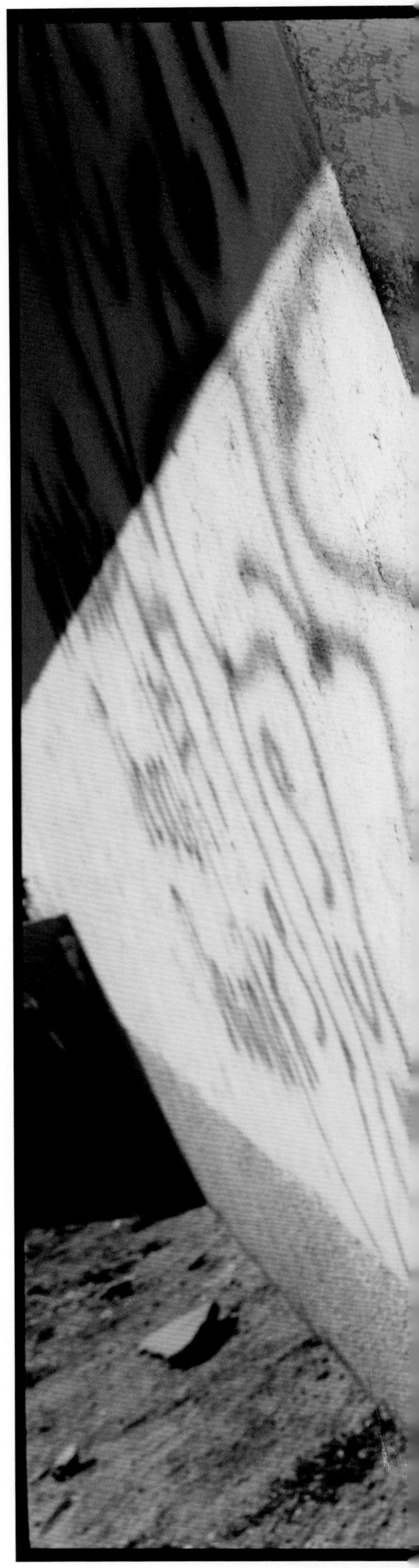

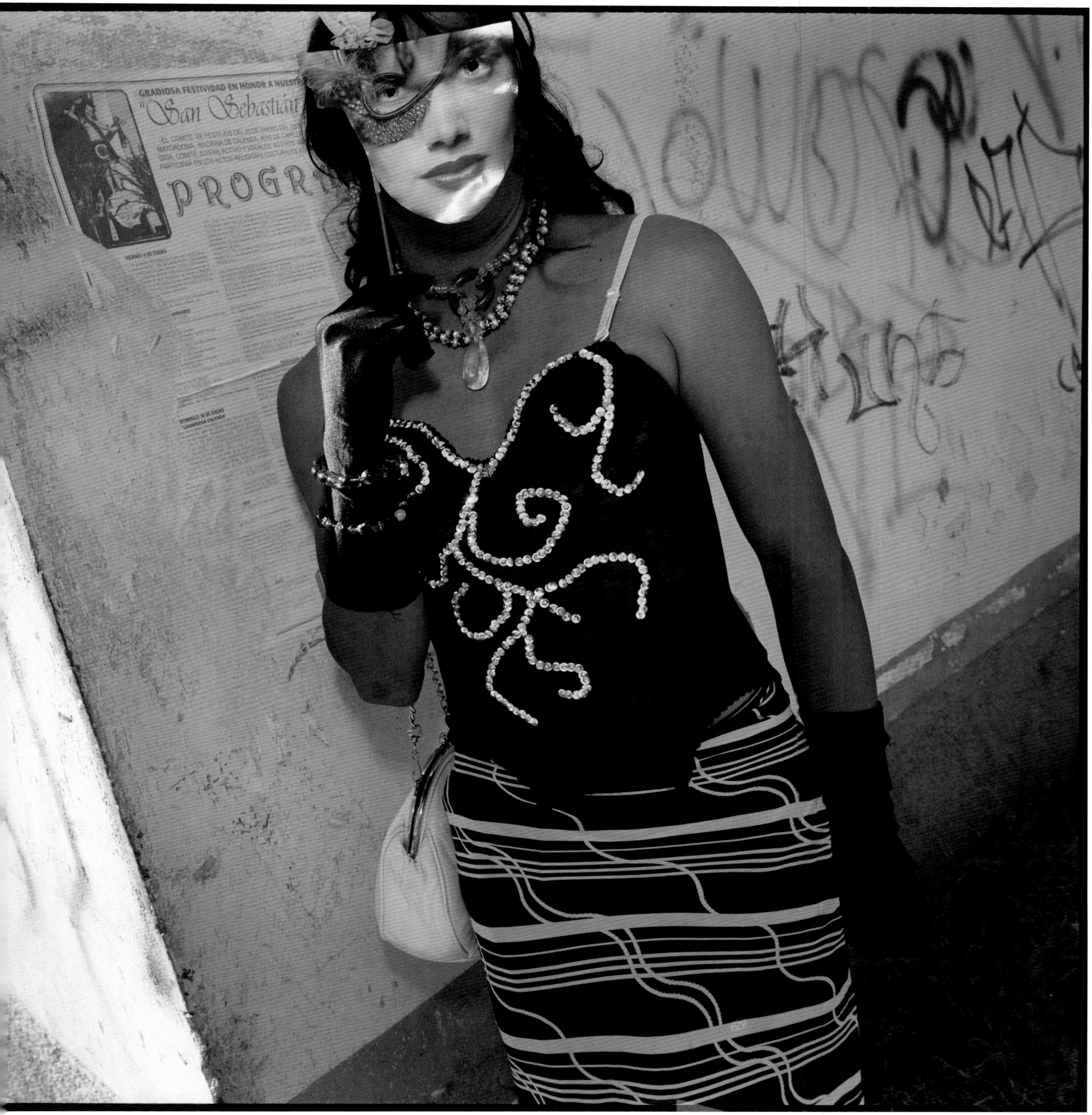
GRADIOSA FESTIVIDAD EN HONOR A
"San Sebastián

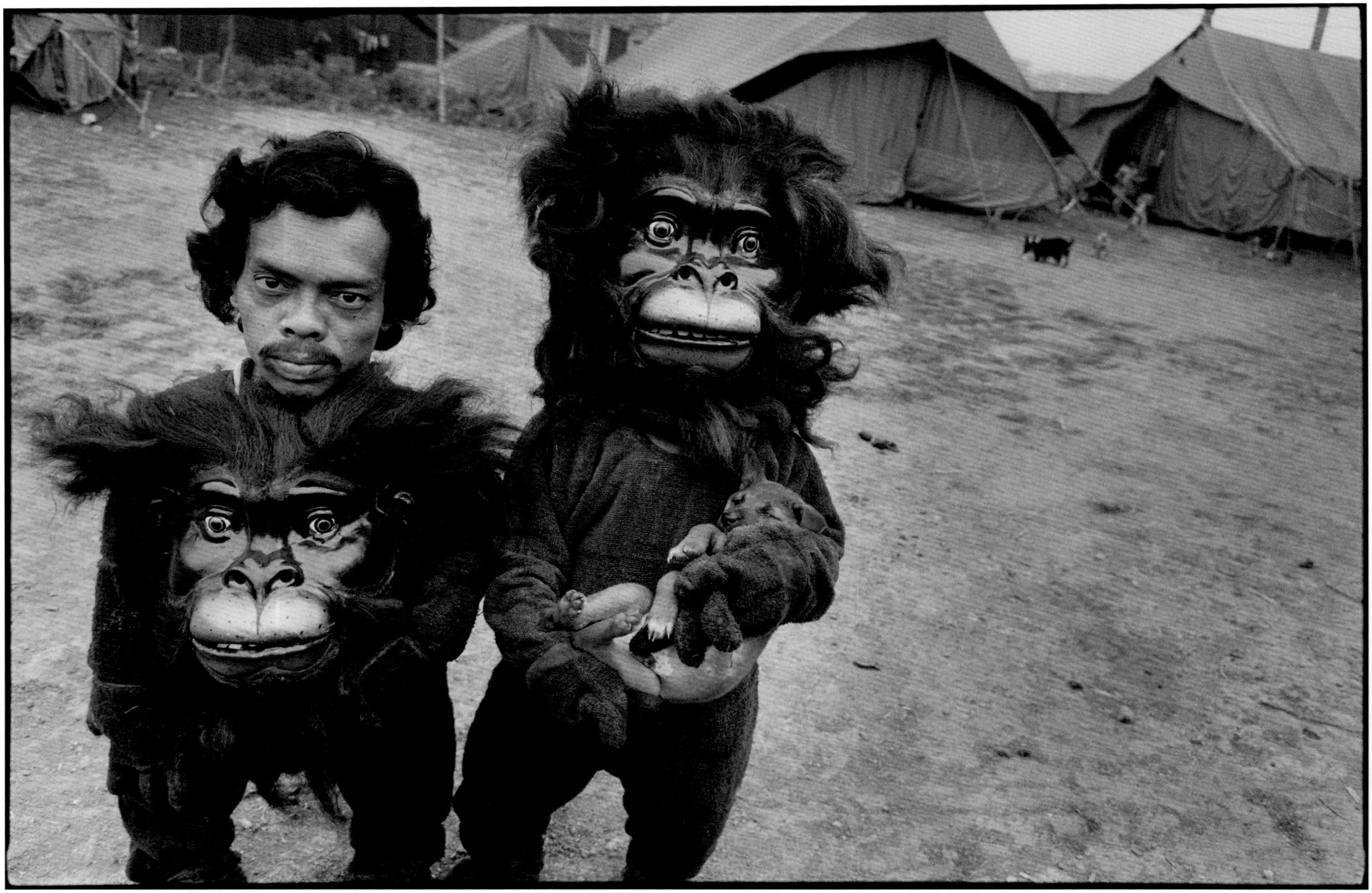

PLATE 86

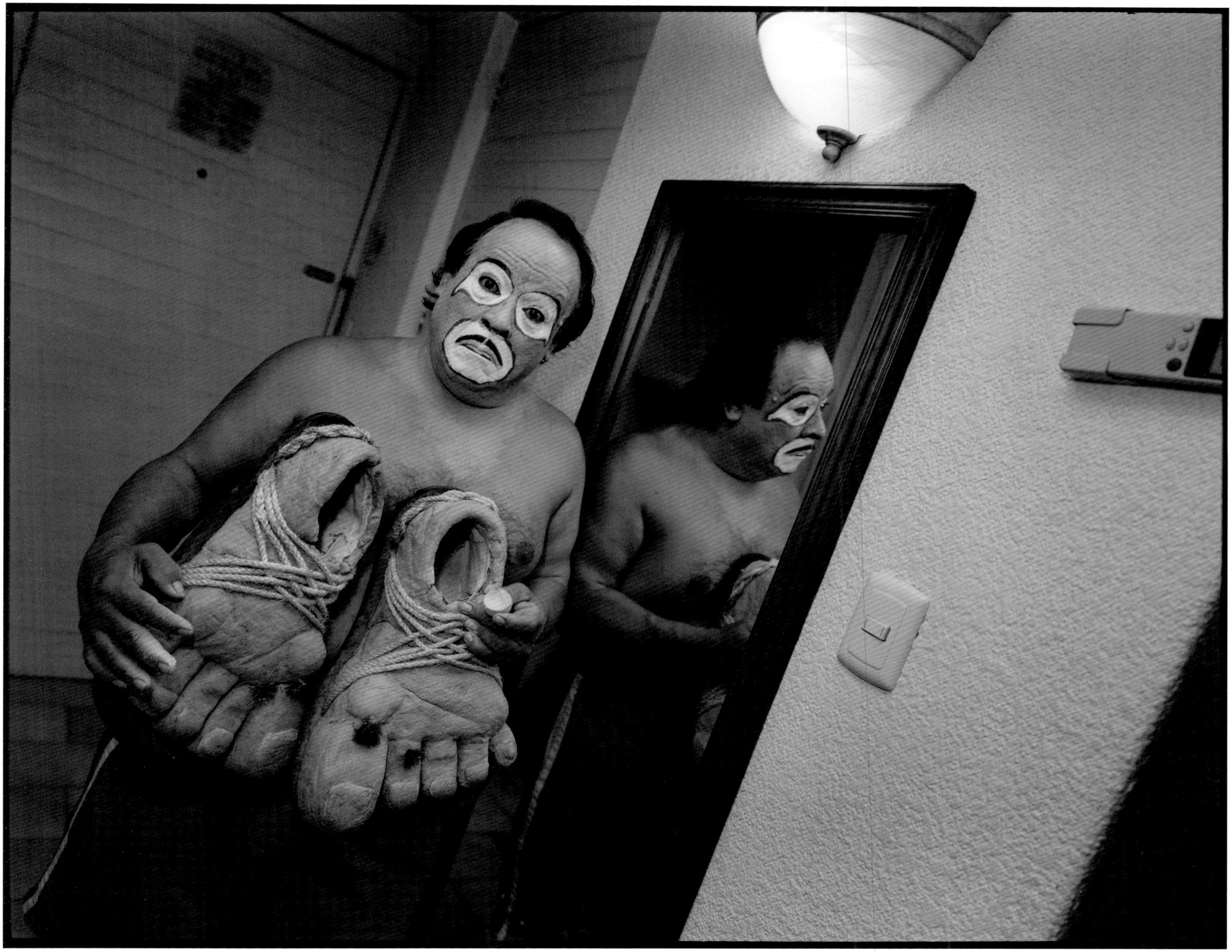

PLATE 87

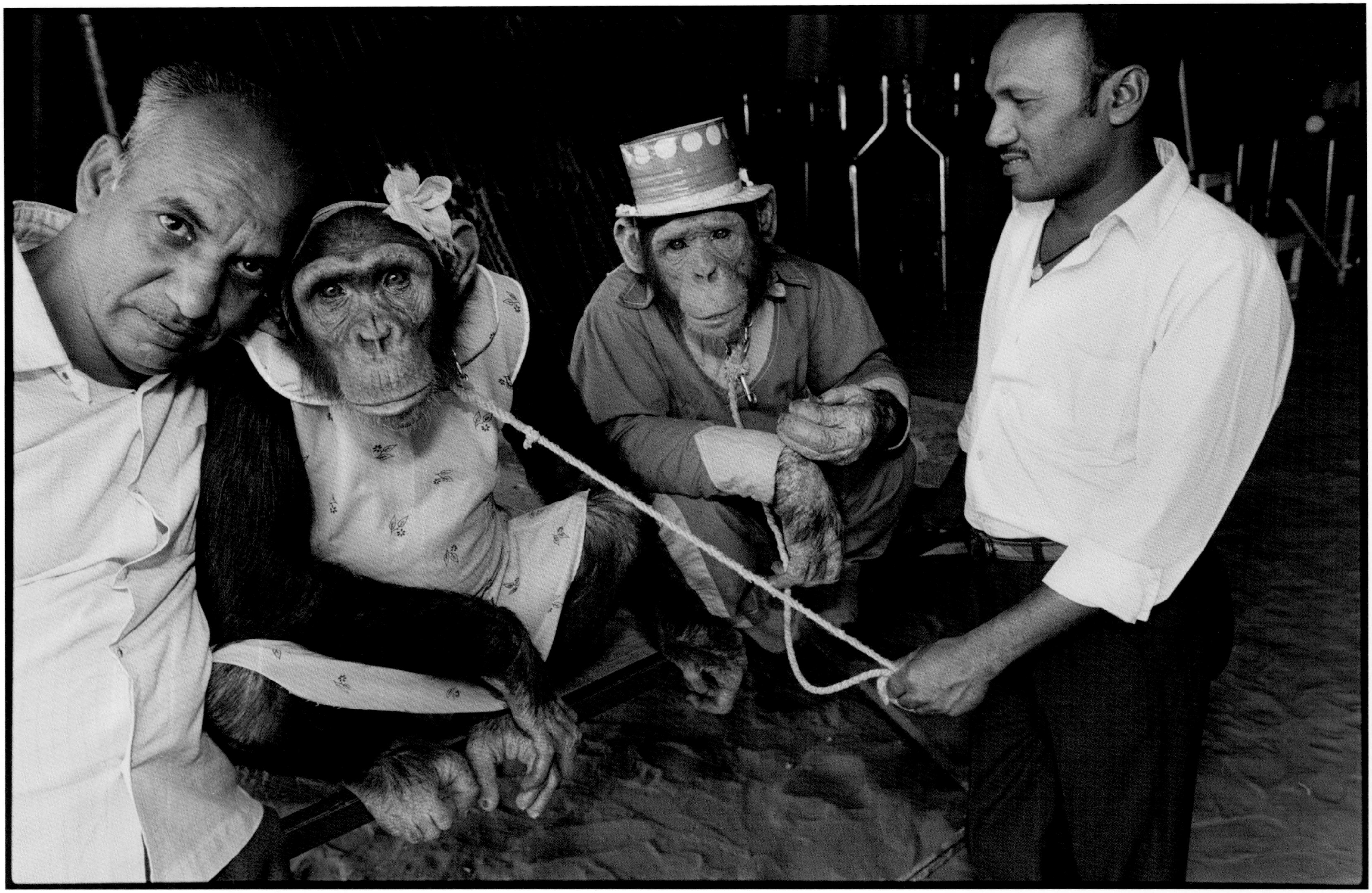

PLATE 88

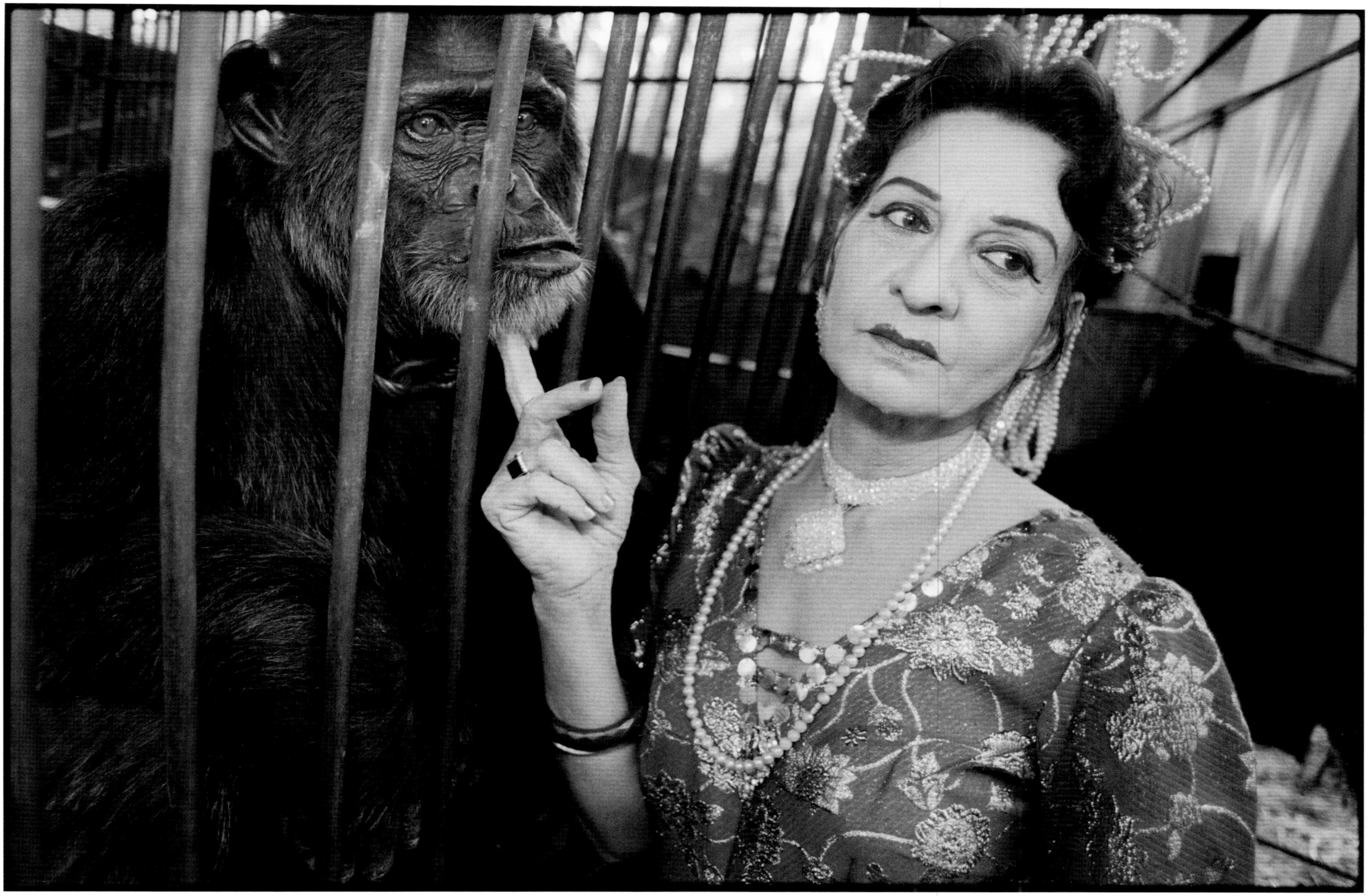

PLATE 89

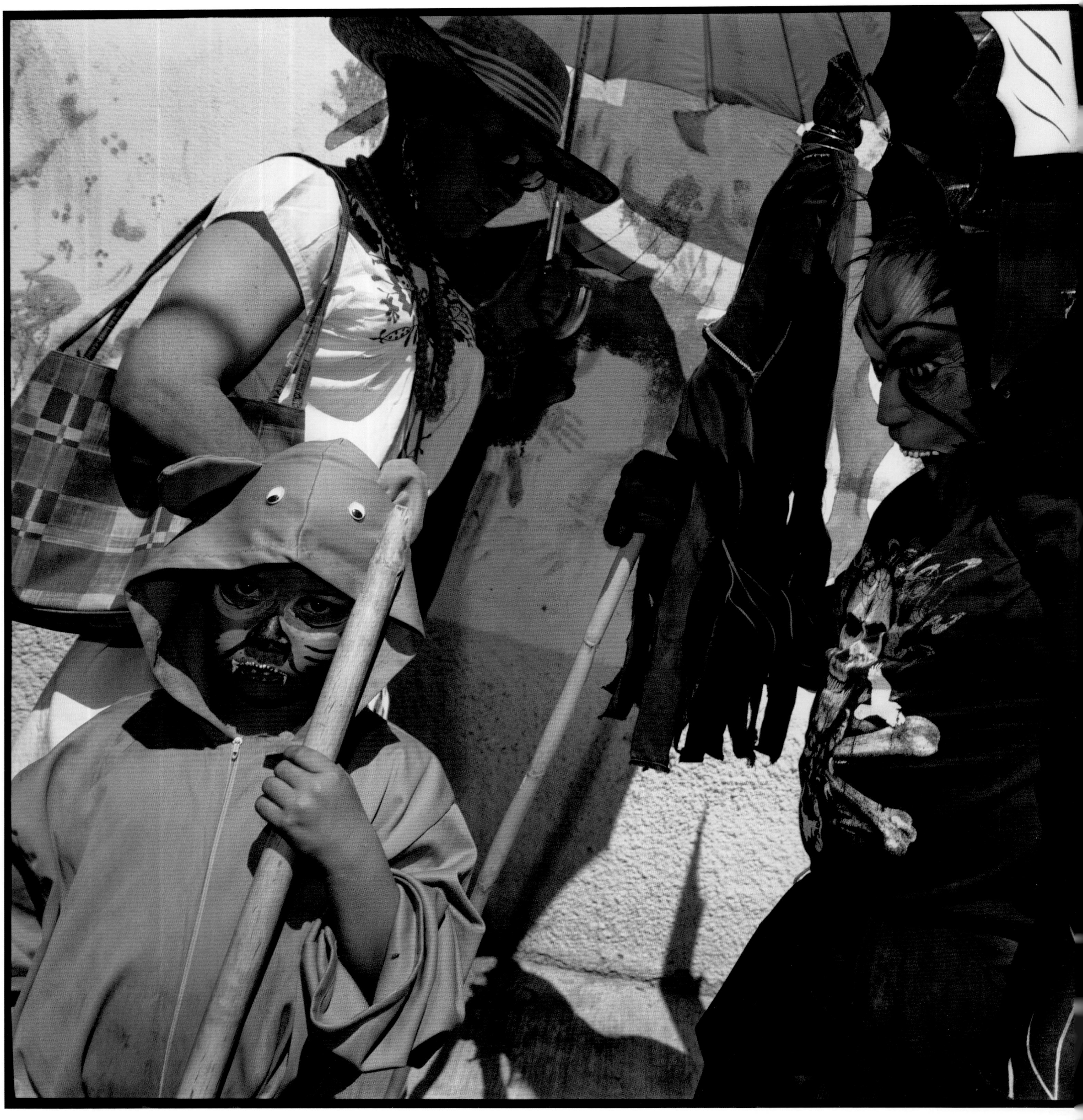

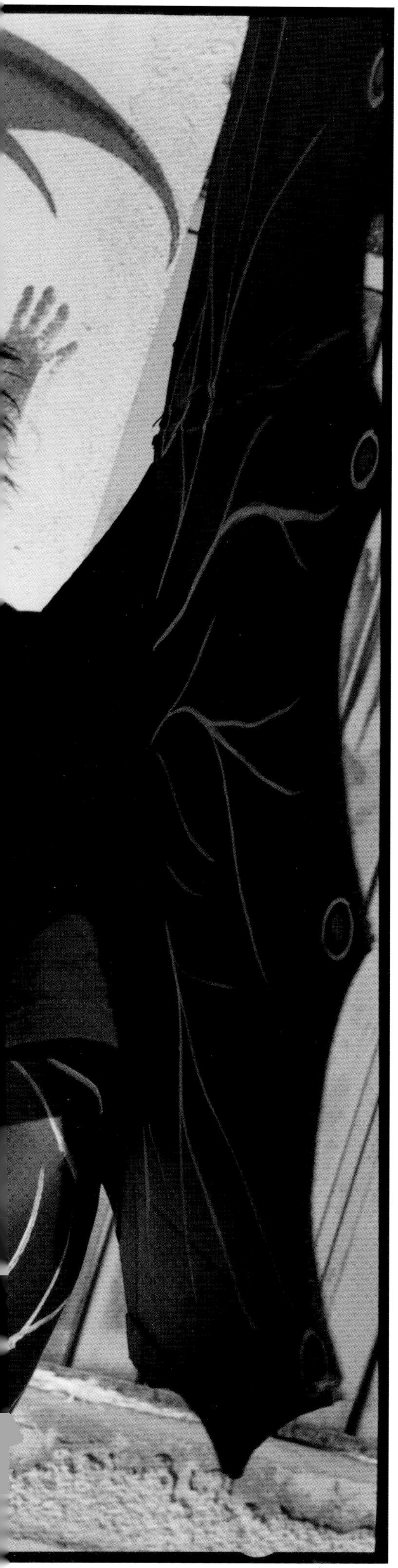

PLATE 90

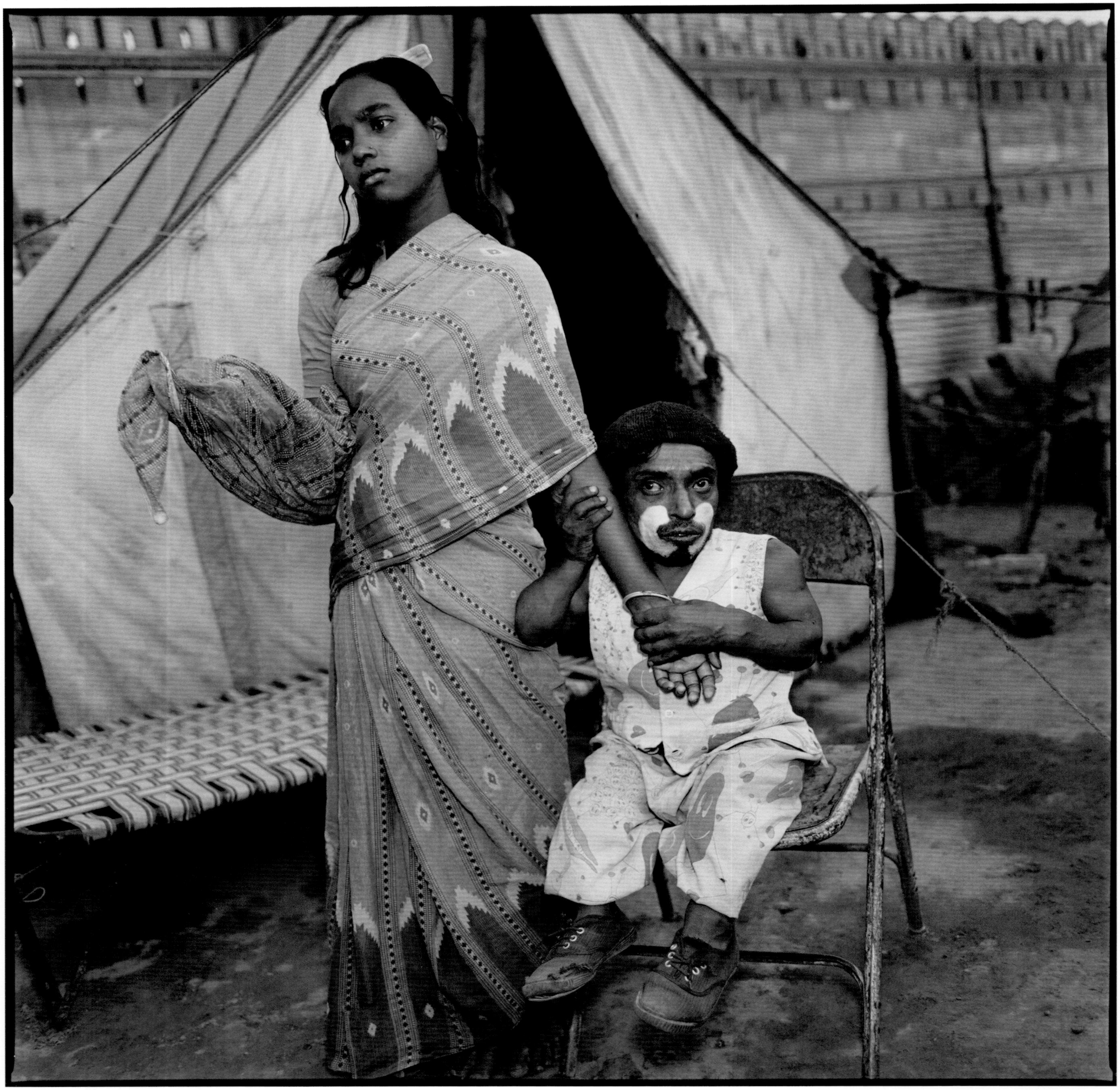

PLATE 91

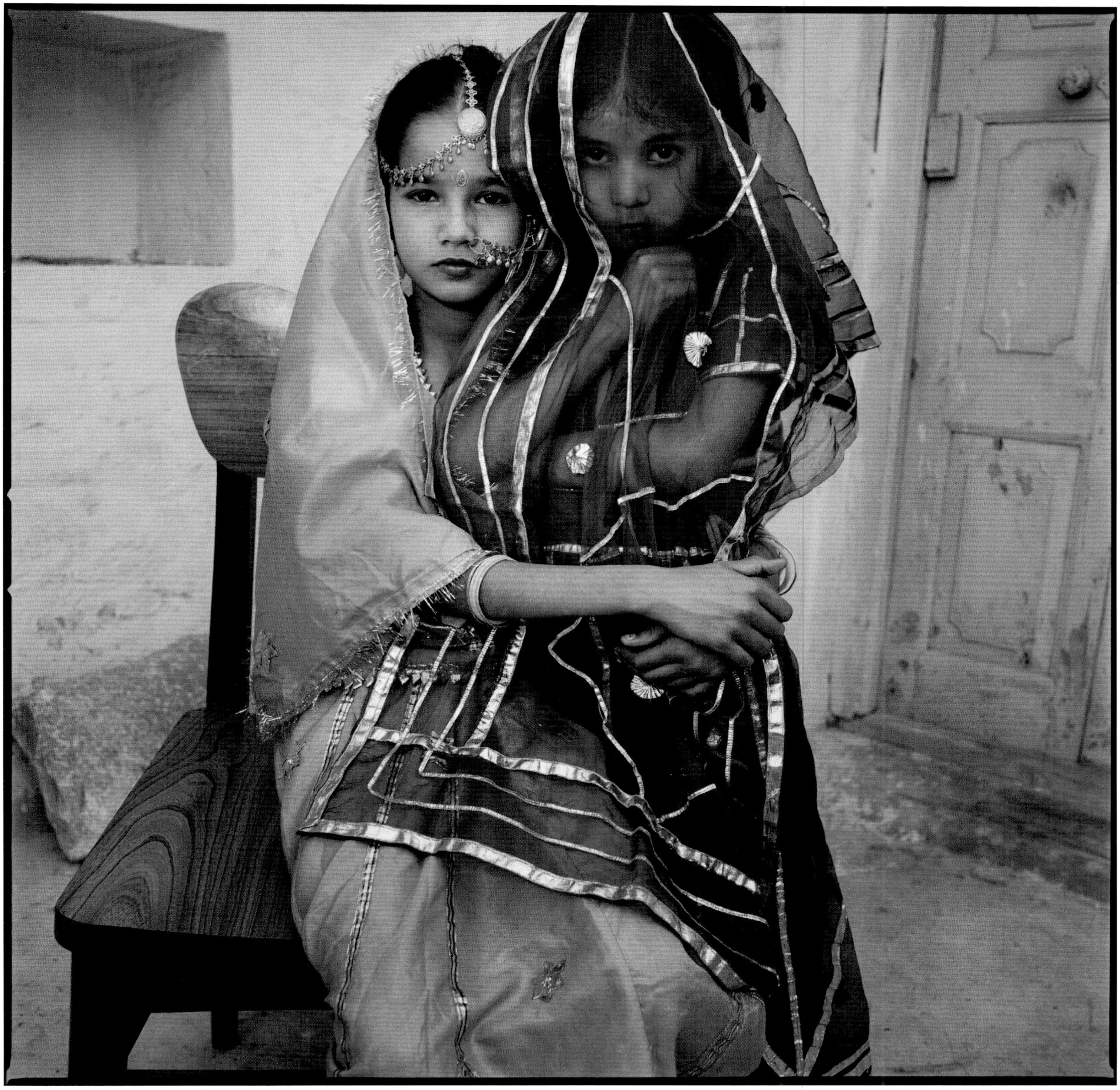

PLATE 92

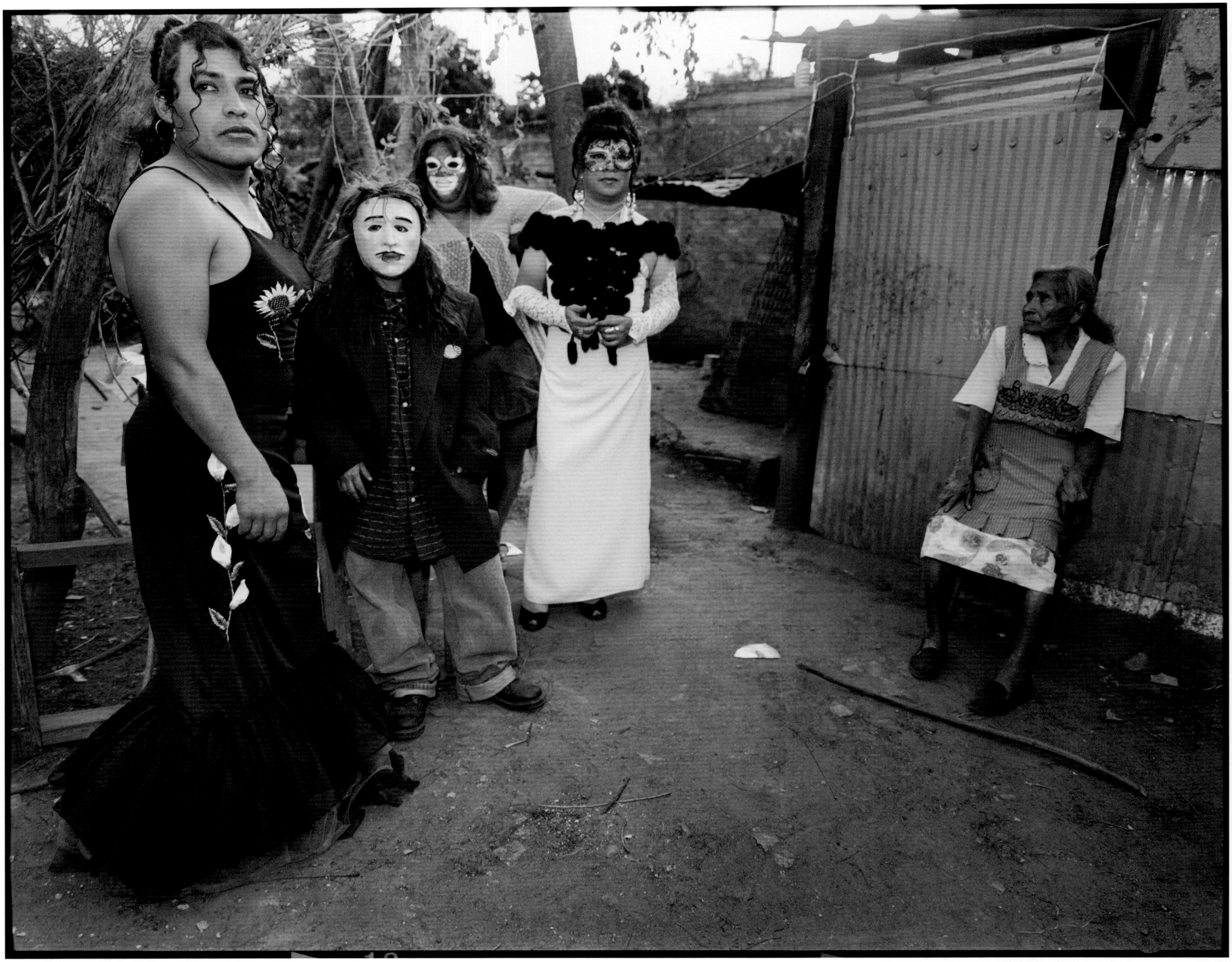

PLATE 93

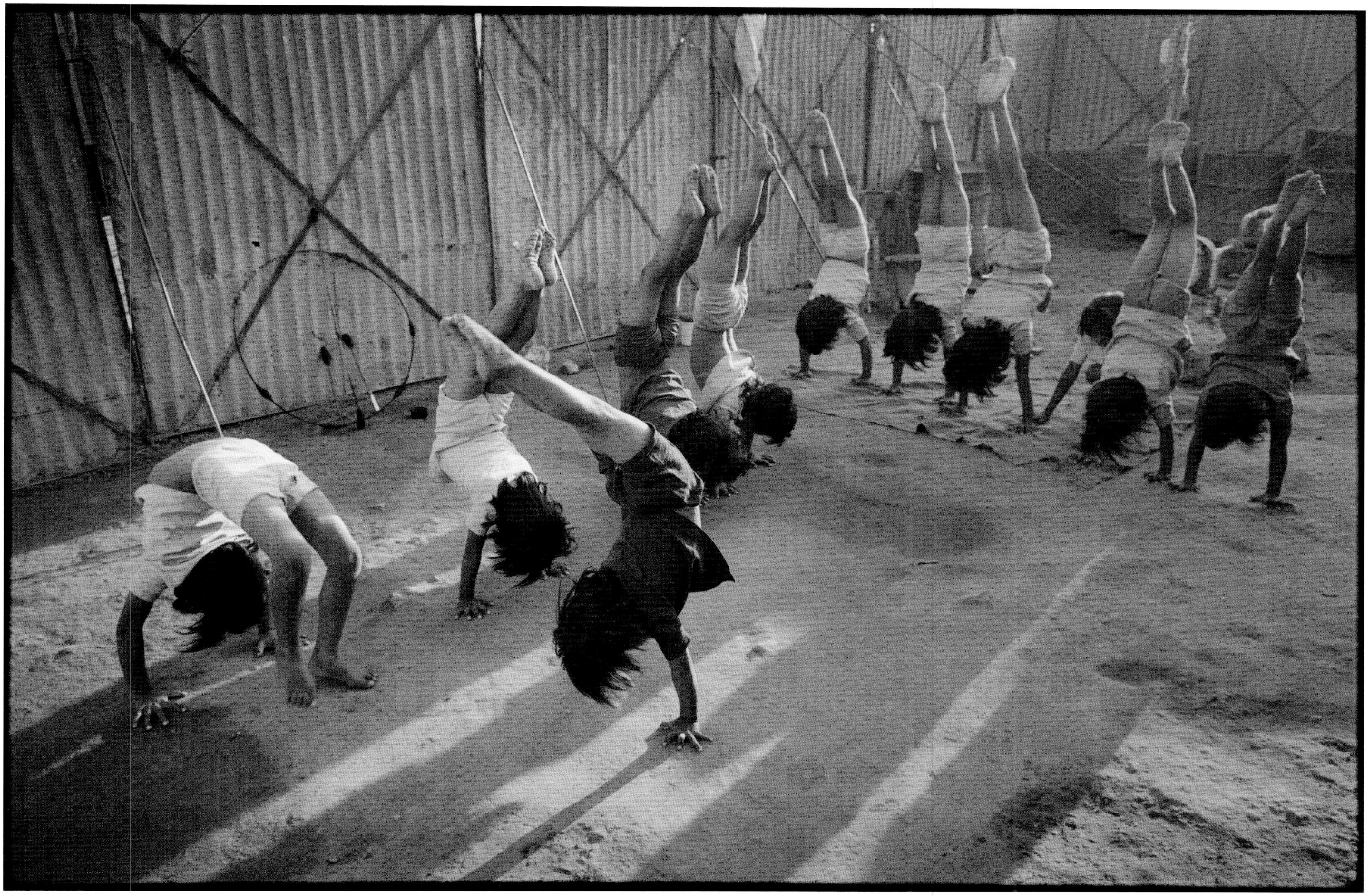

PLATE 94

PLATE 95

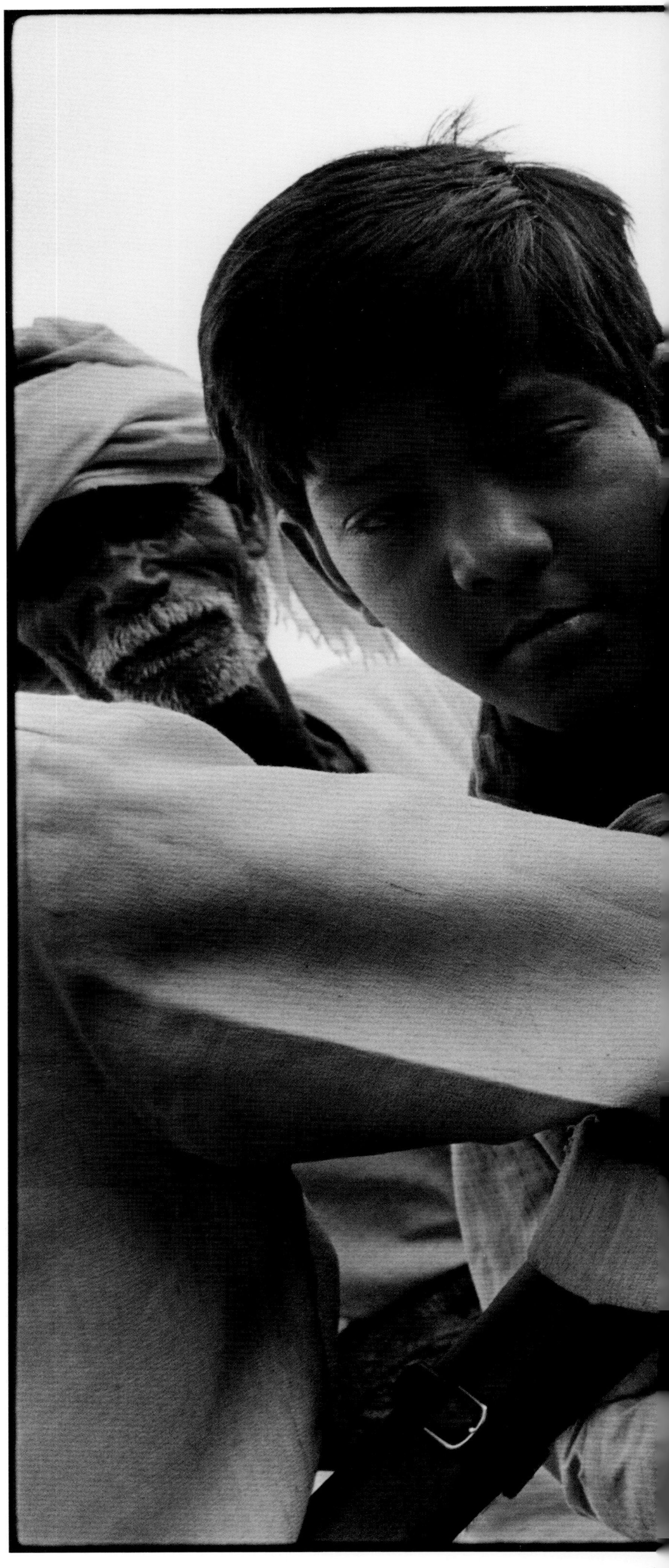

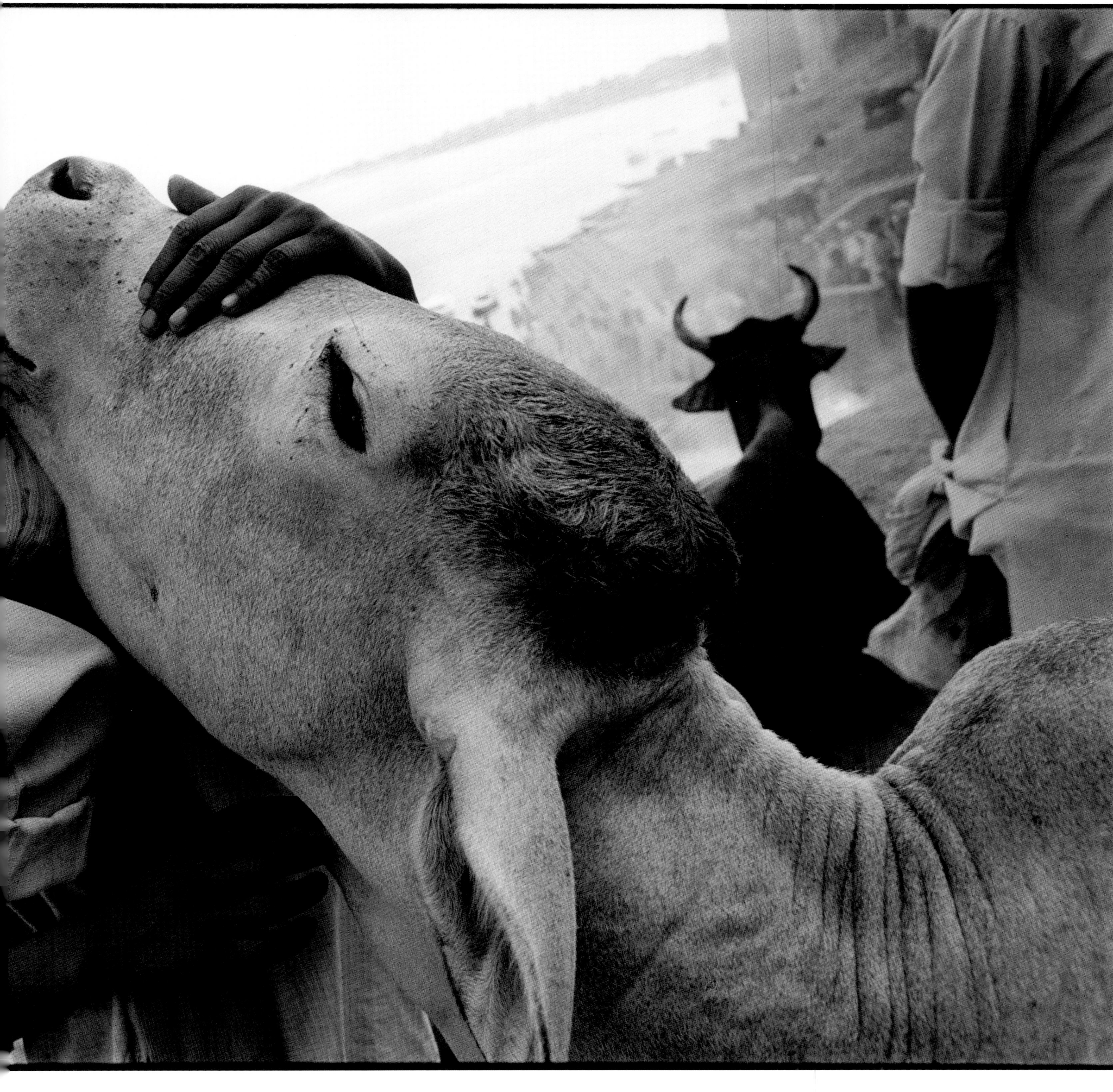

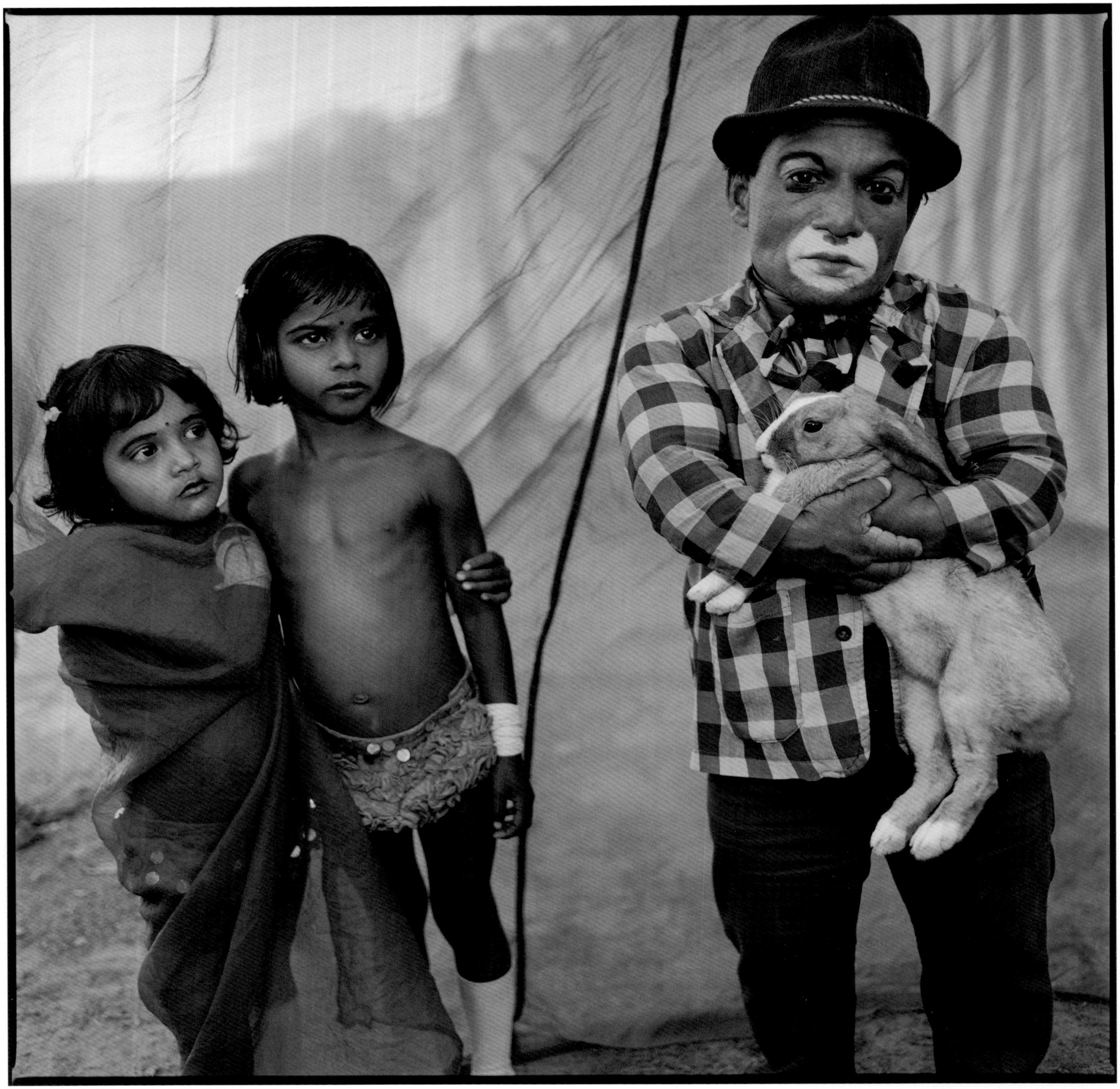

PLATE 96

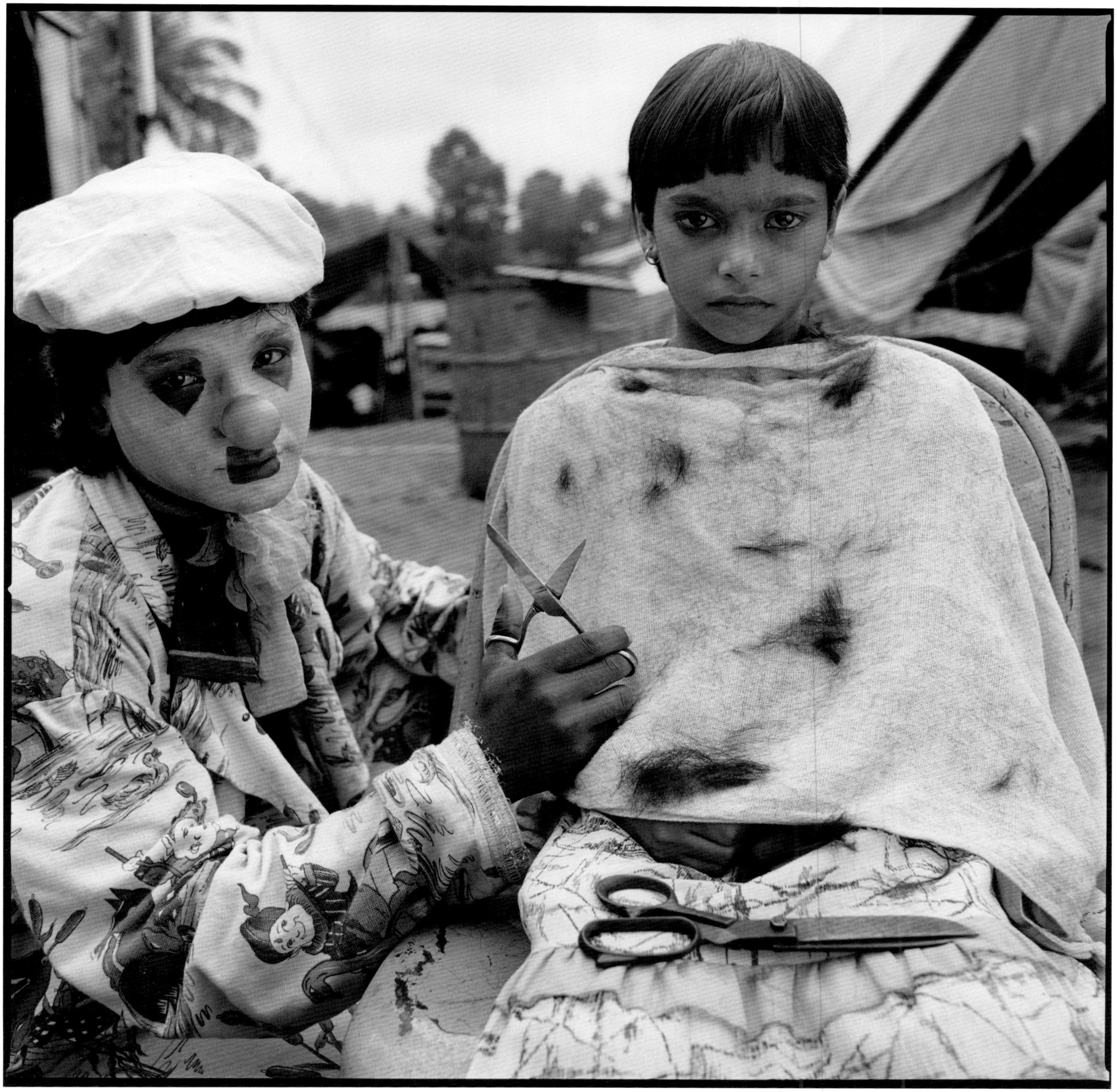

PLATE 97

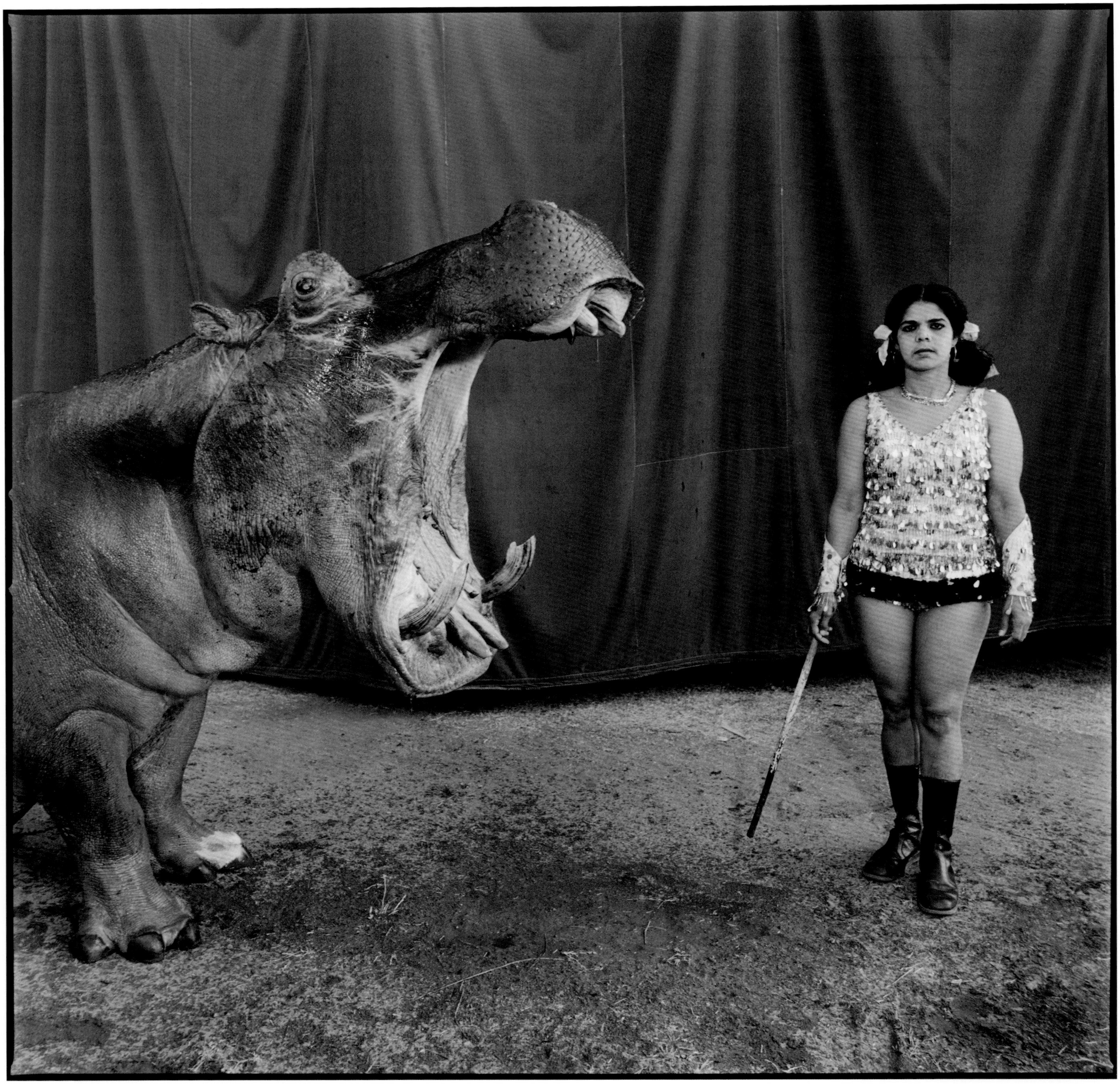

PLATE 98

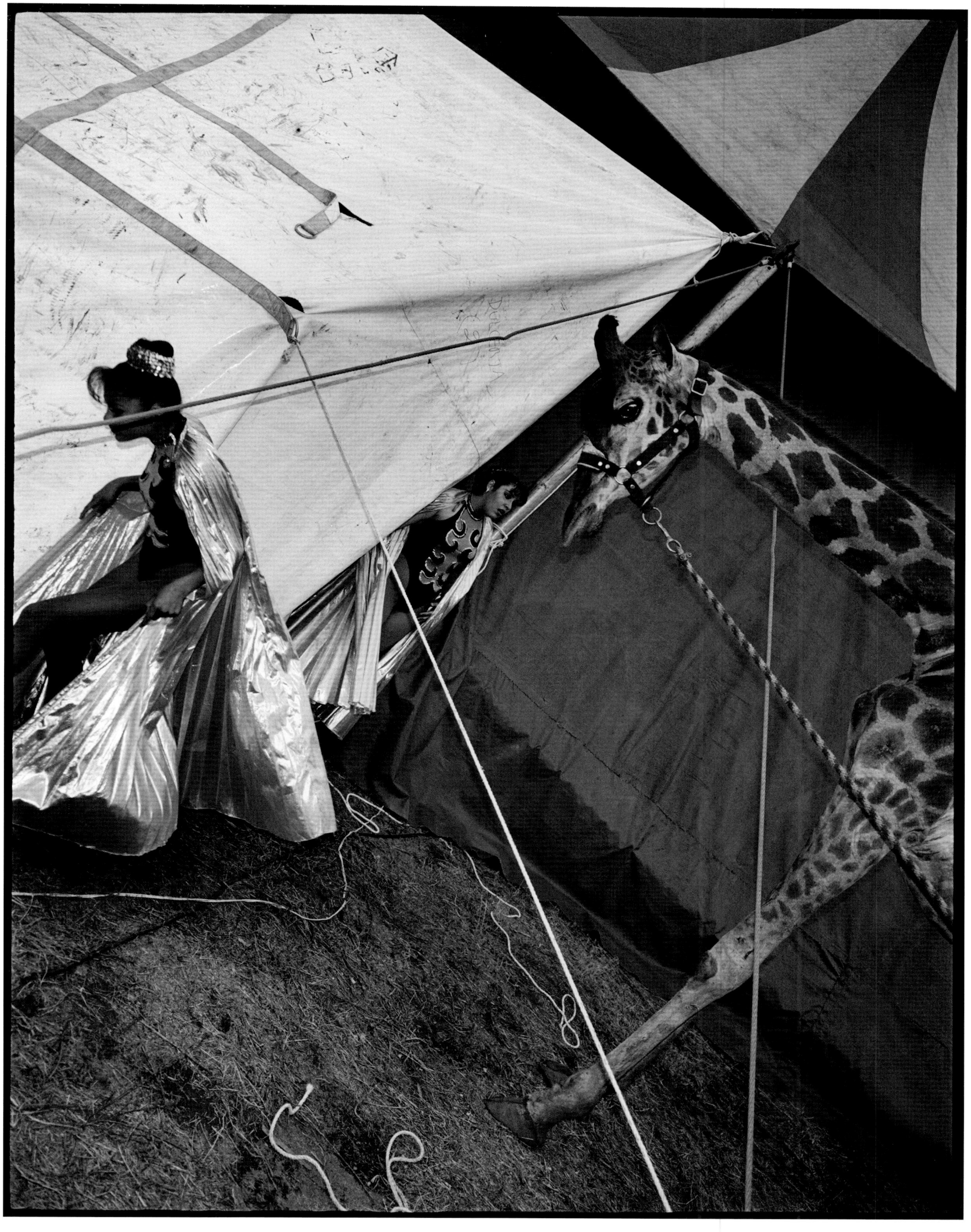

PLATE 99

PLATE 100

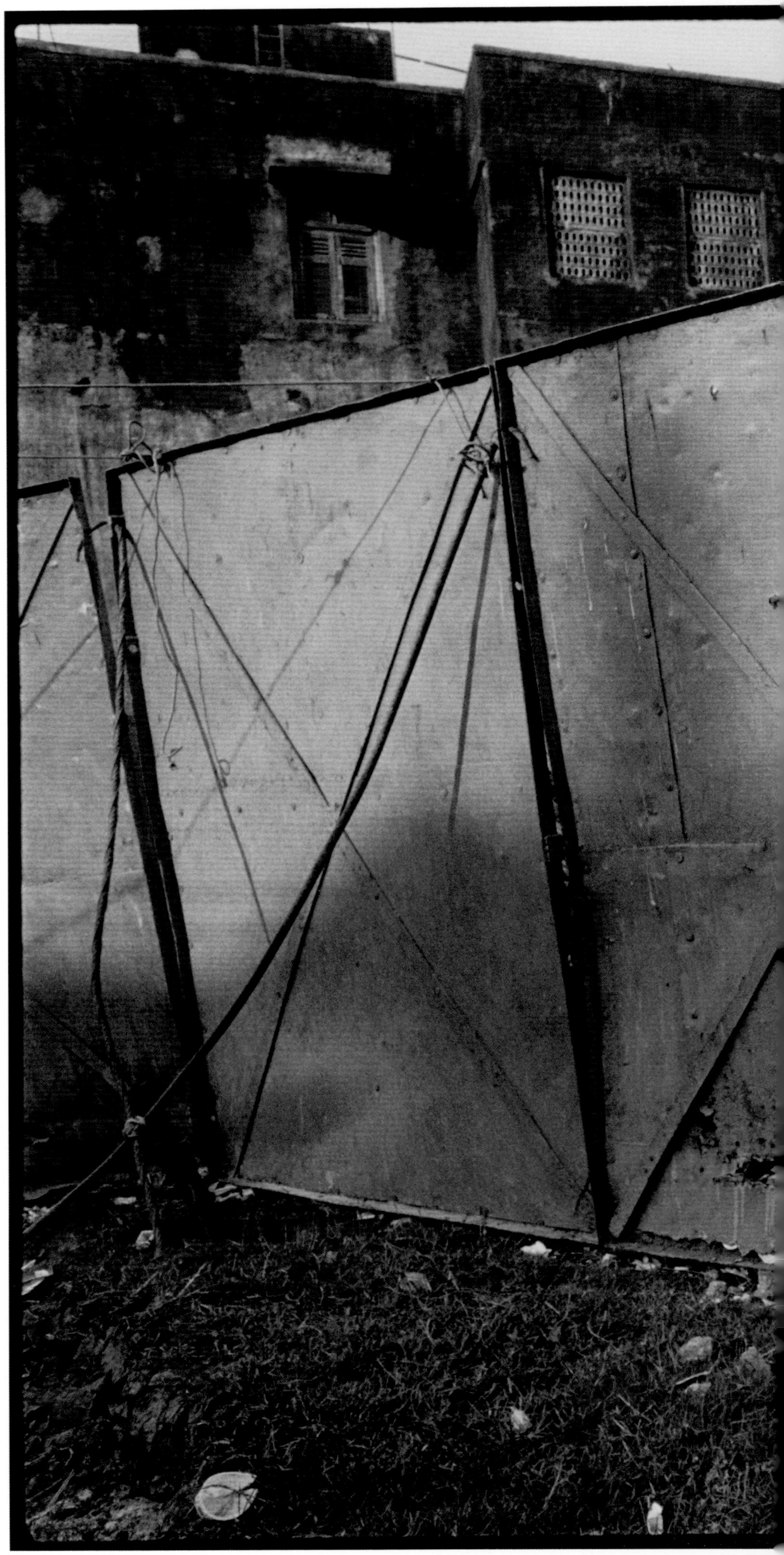

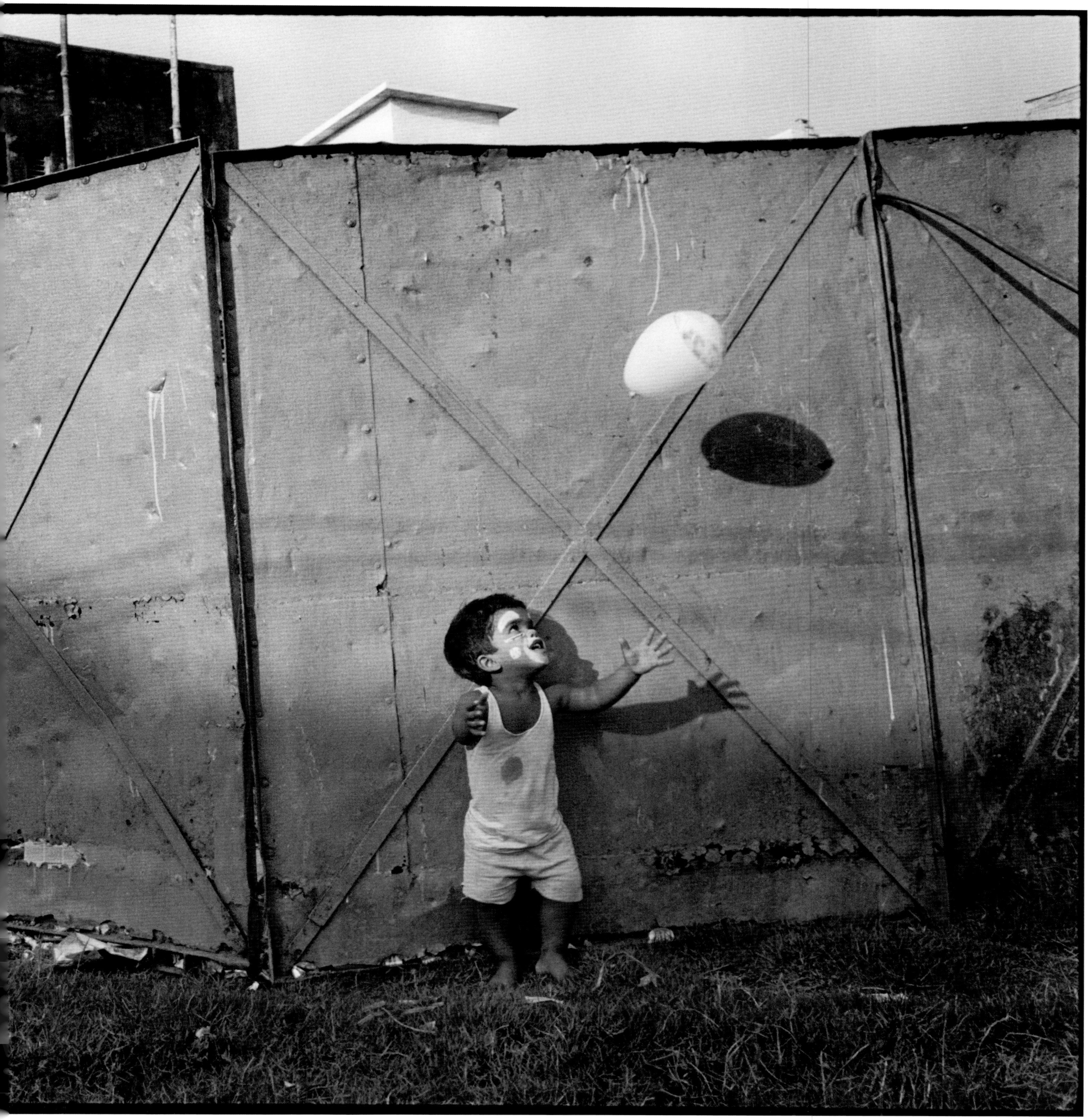

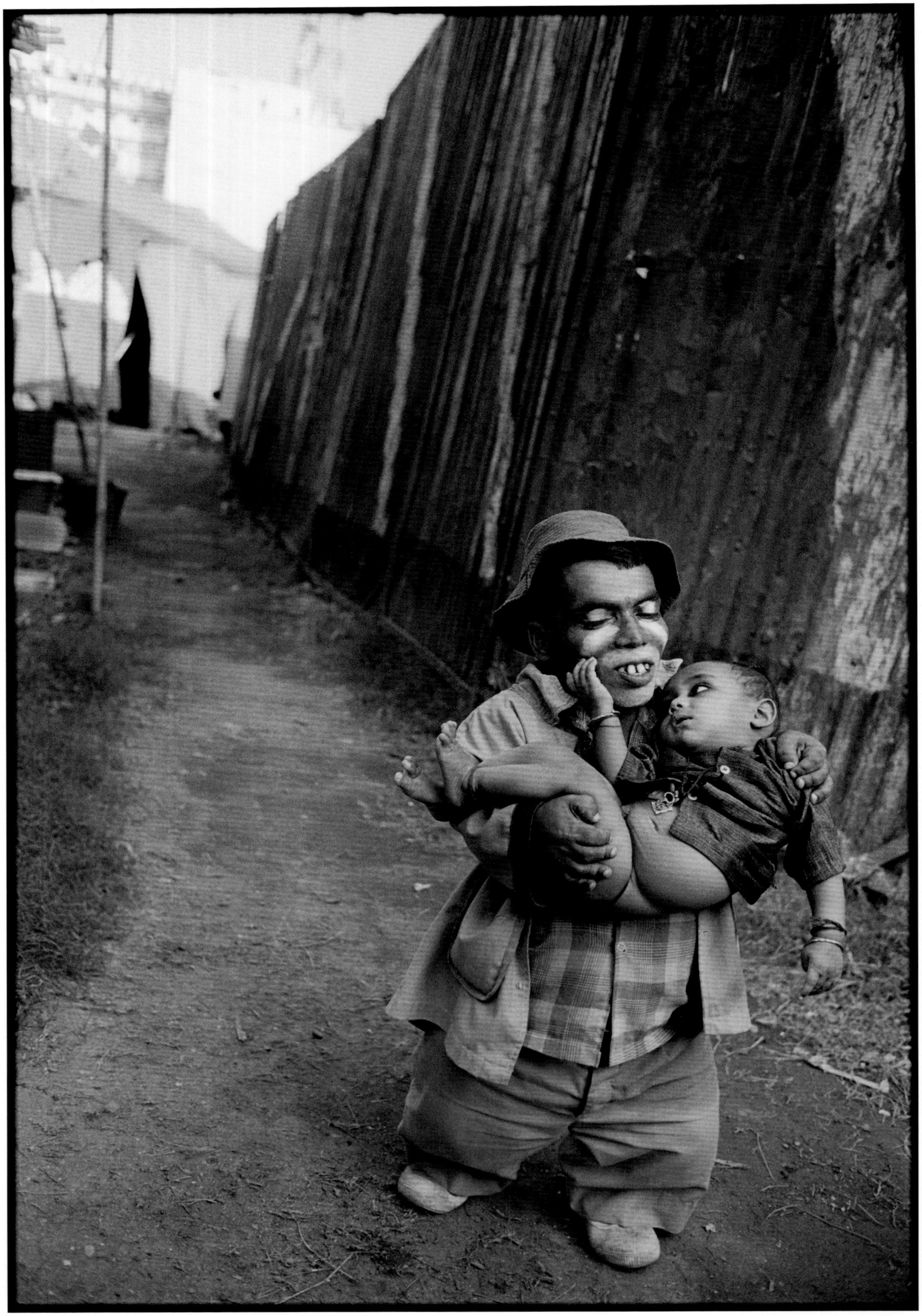

PLATE 101

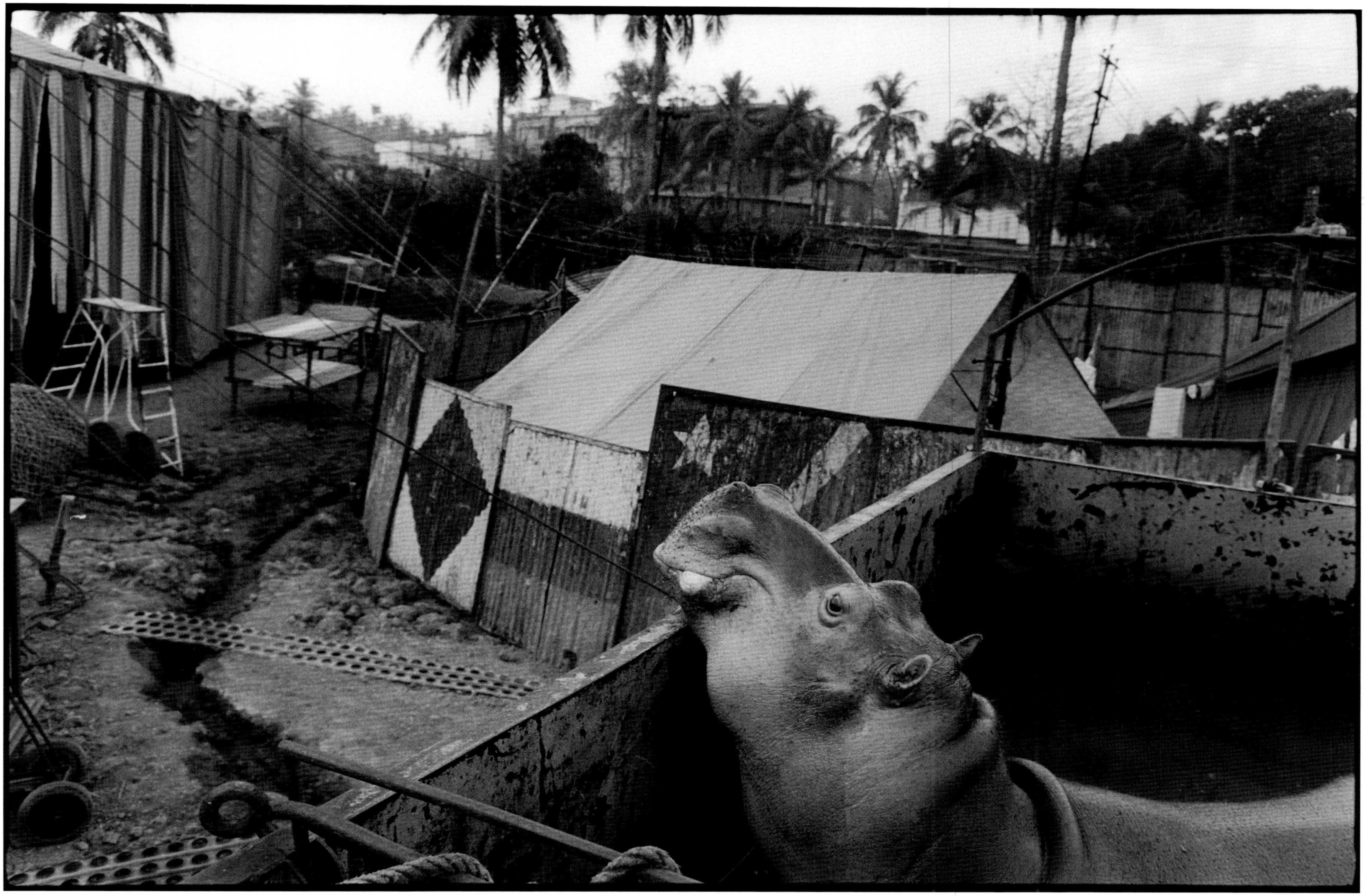

PLATE 102

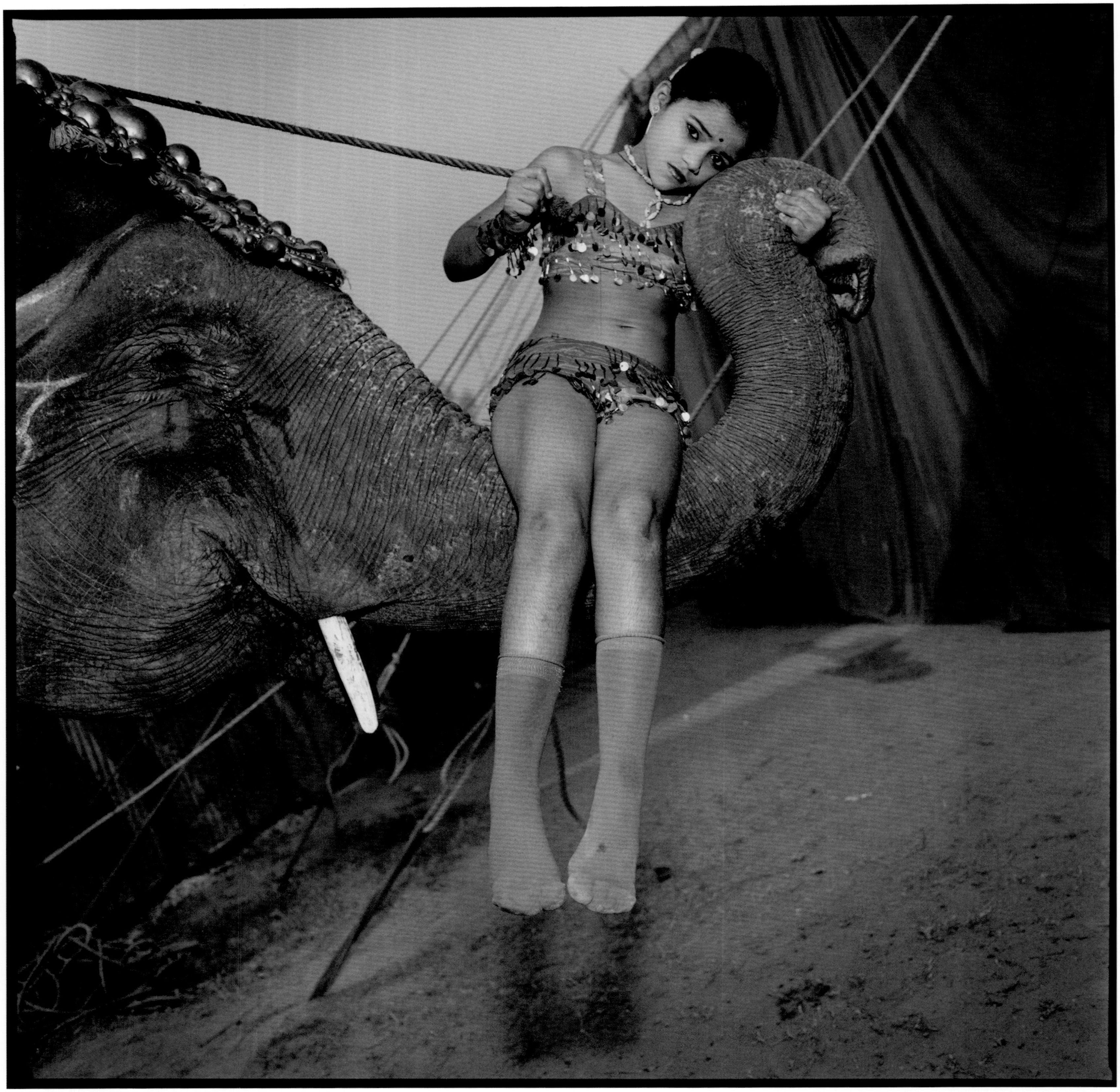

PLATE 103

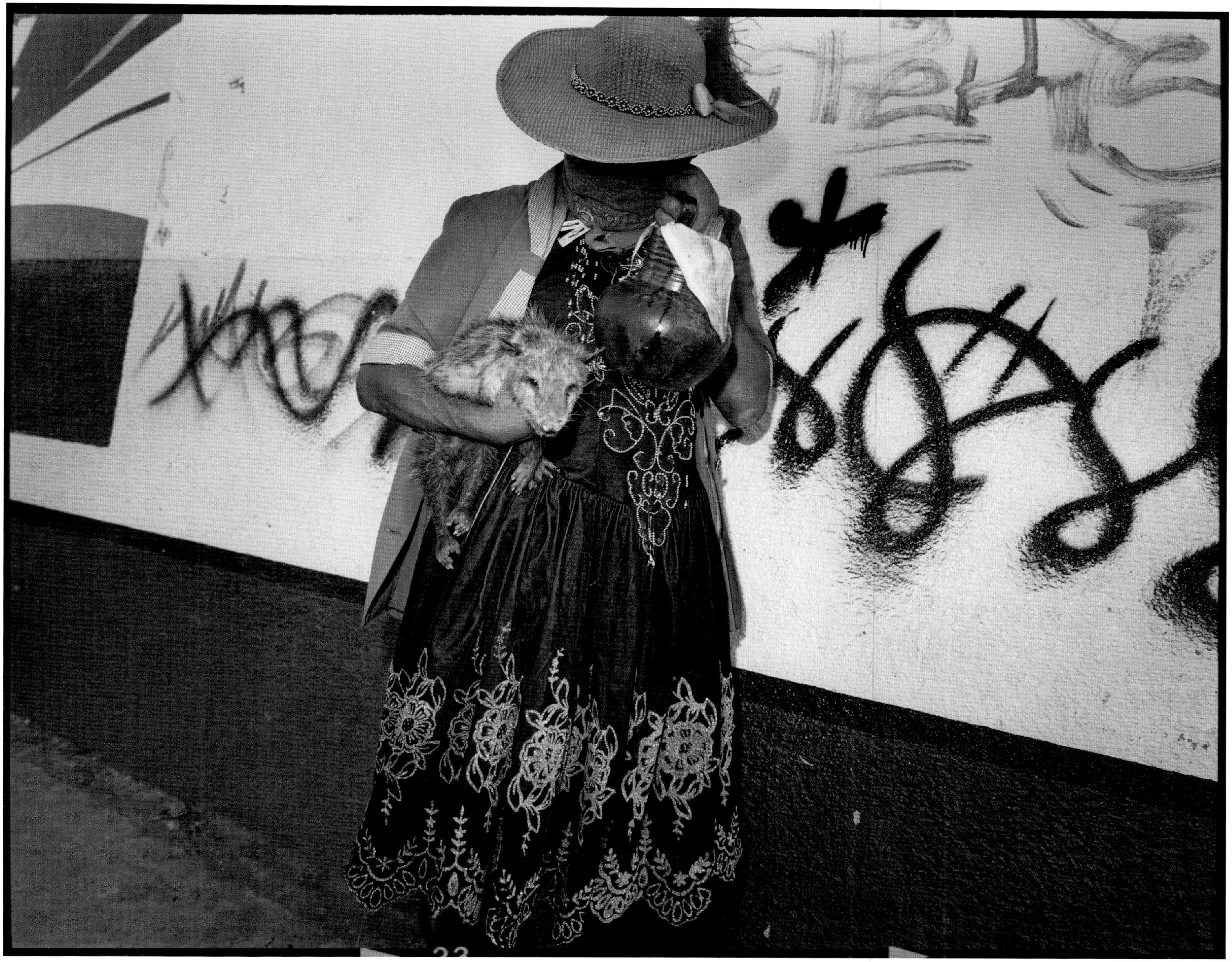

PLATE 104

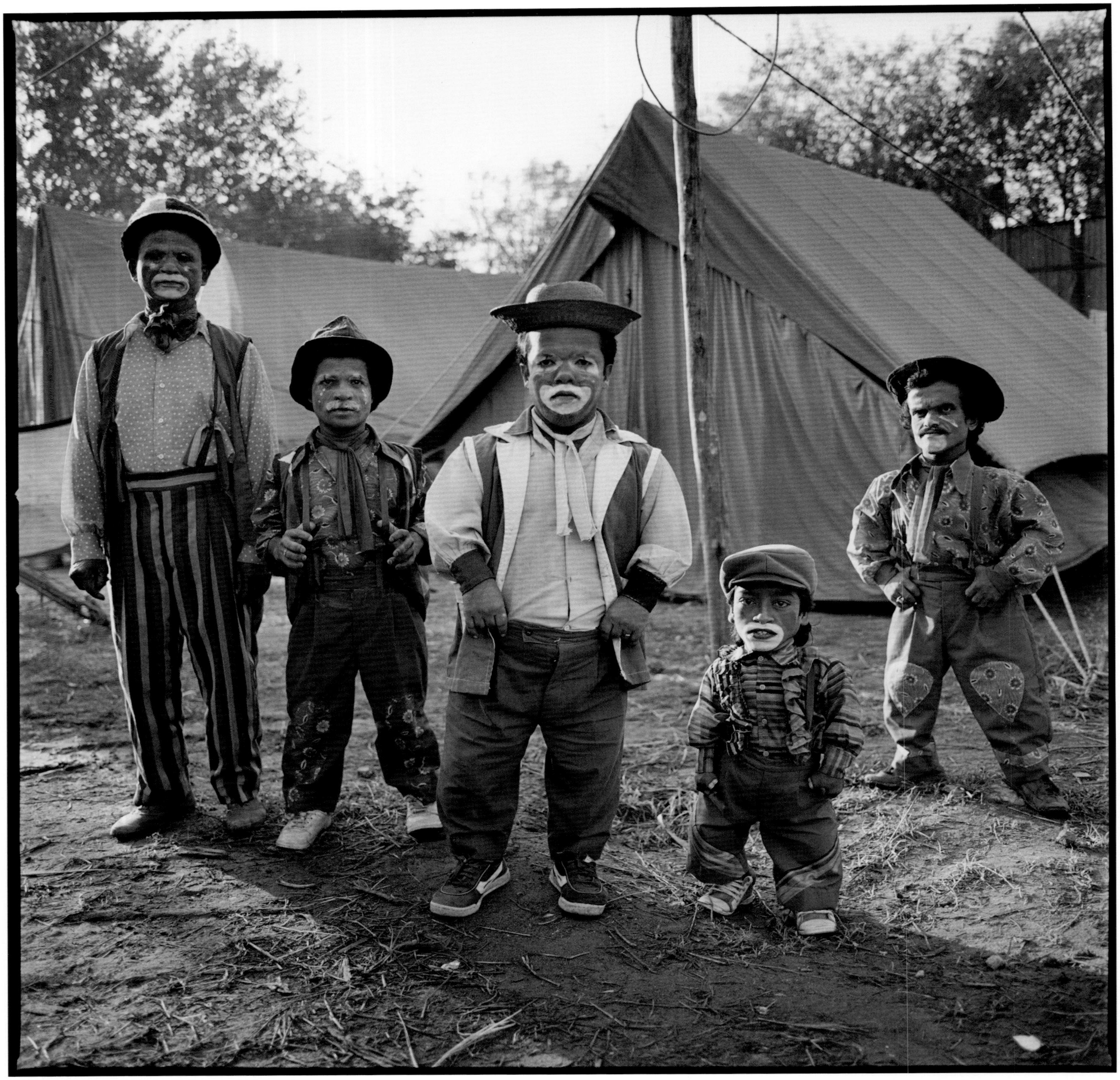

PLATE 105

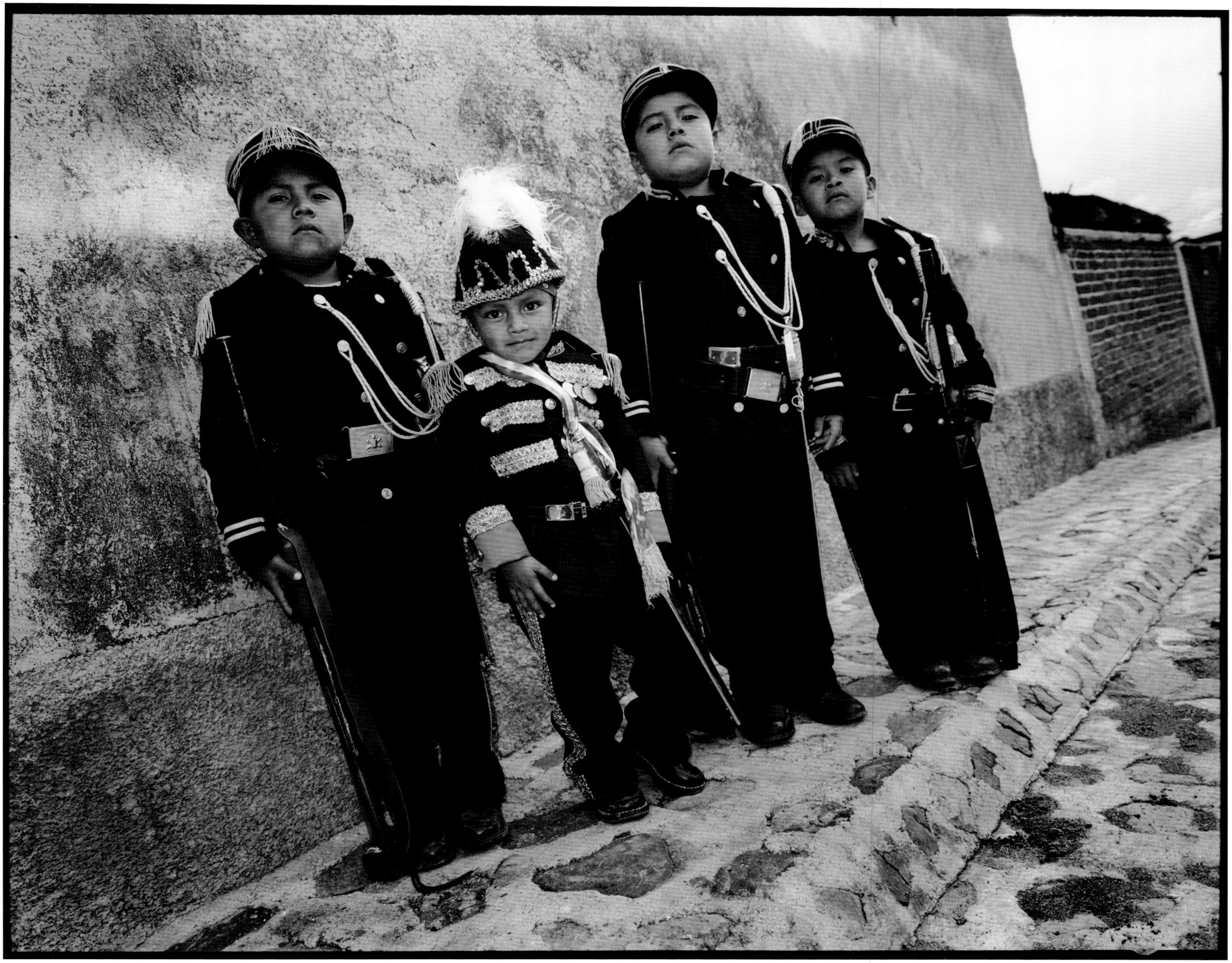

PLATE 106

PLATE 107

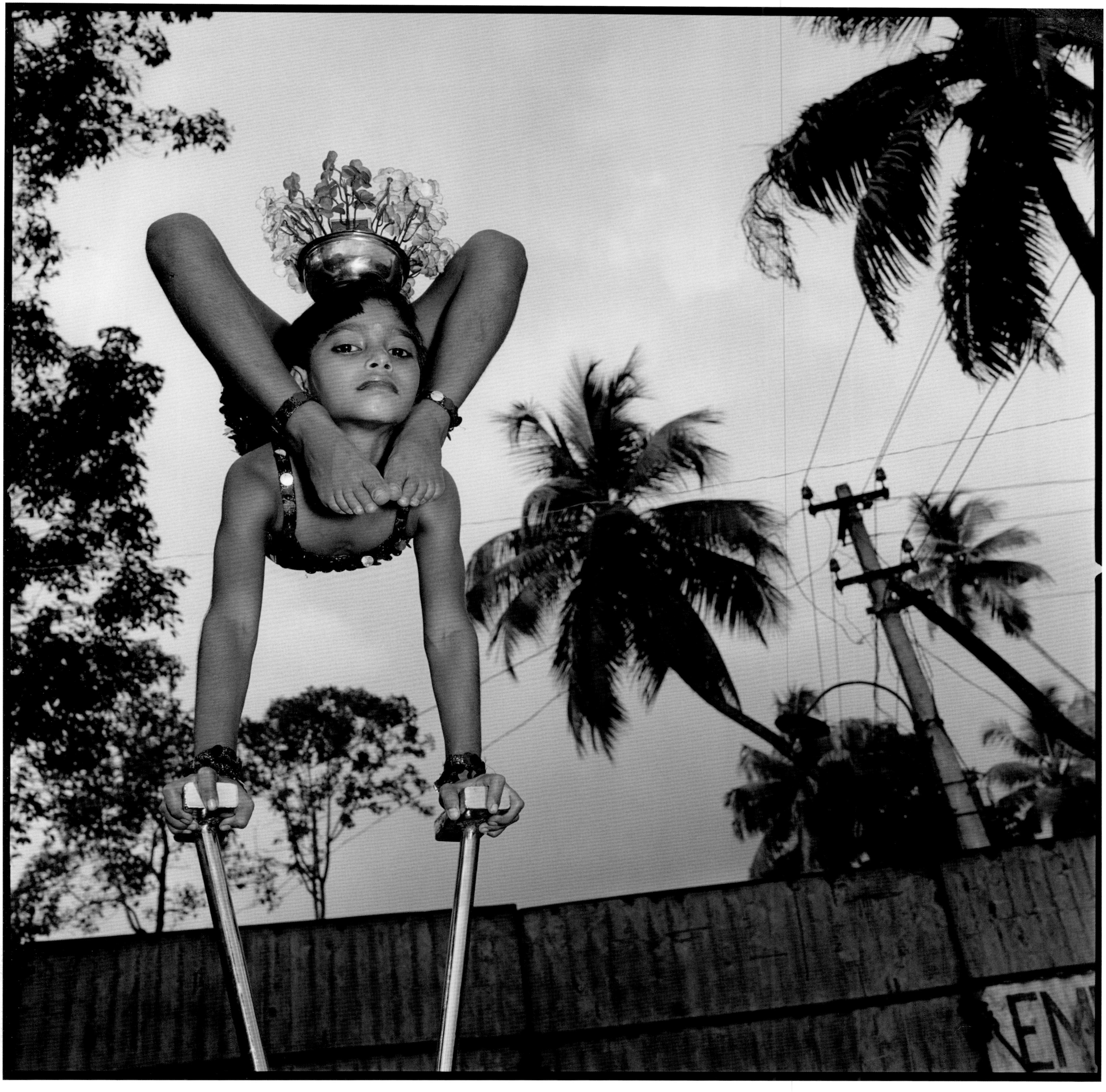

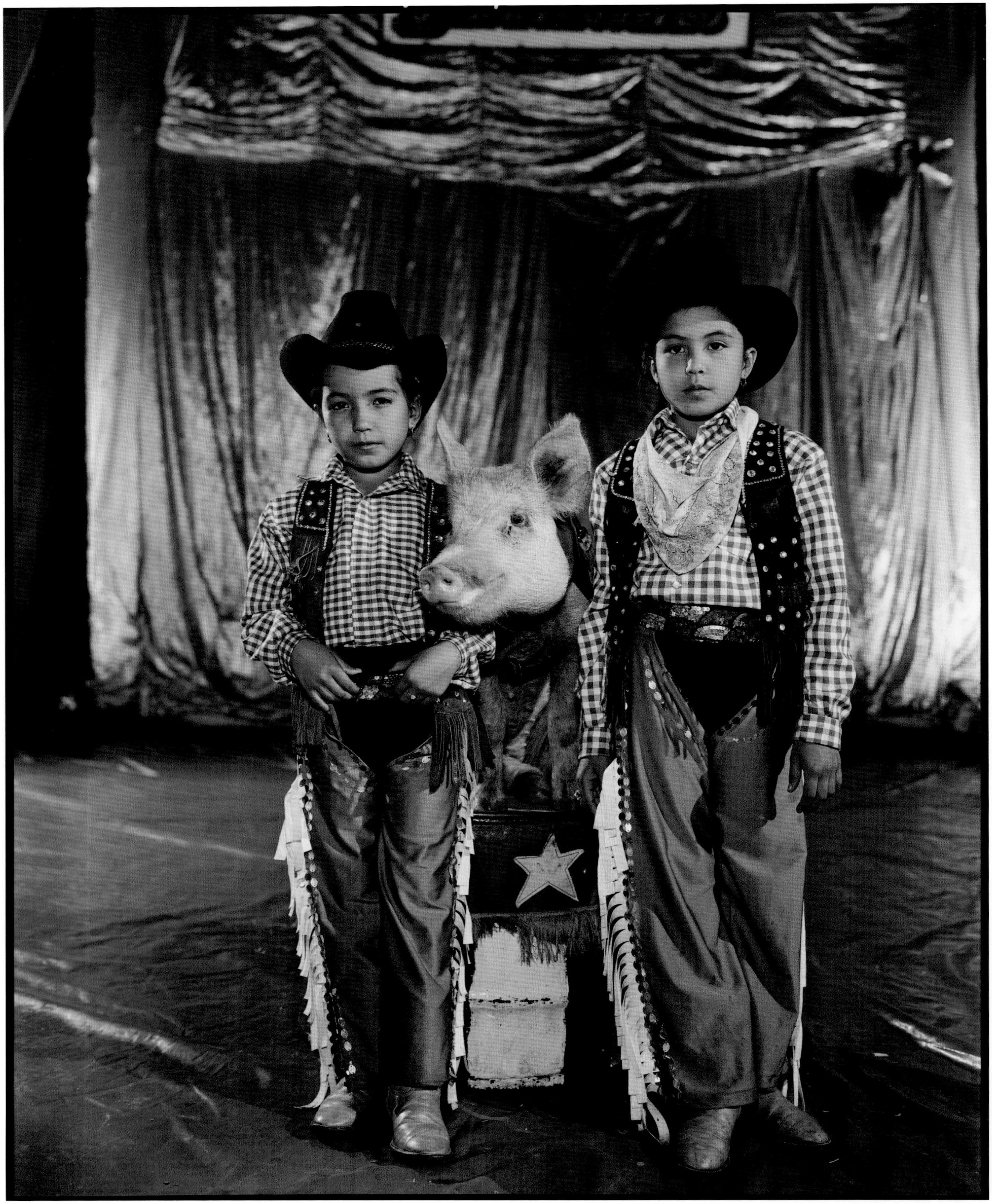

PLATE 108

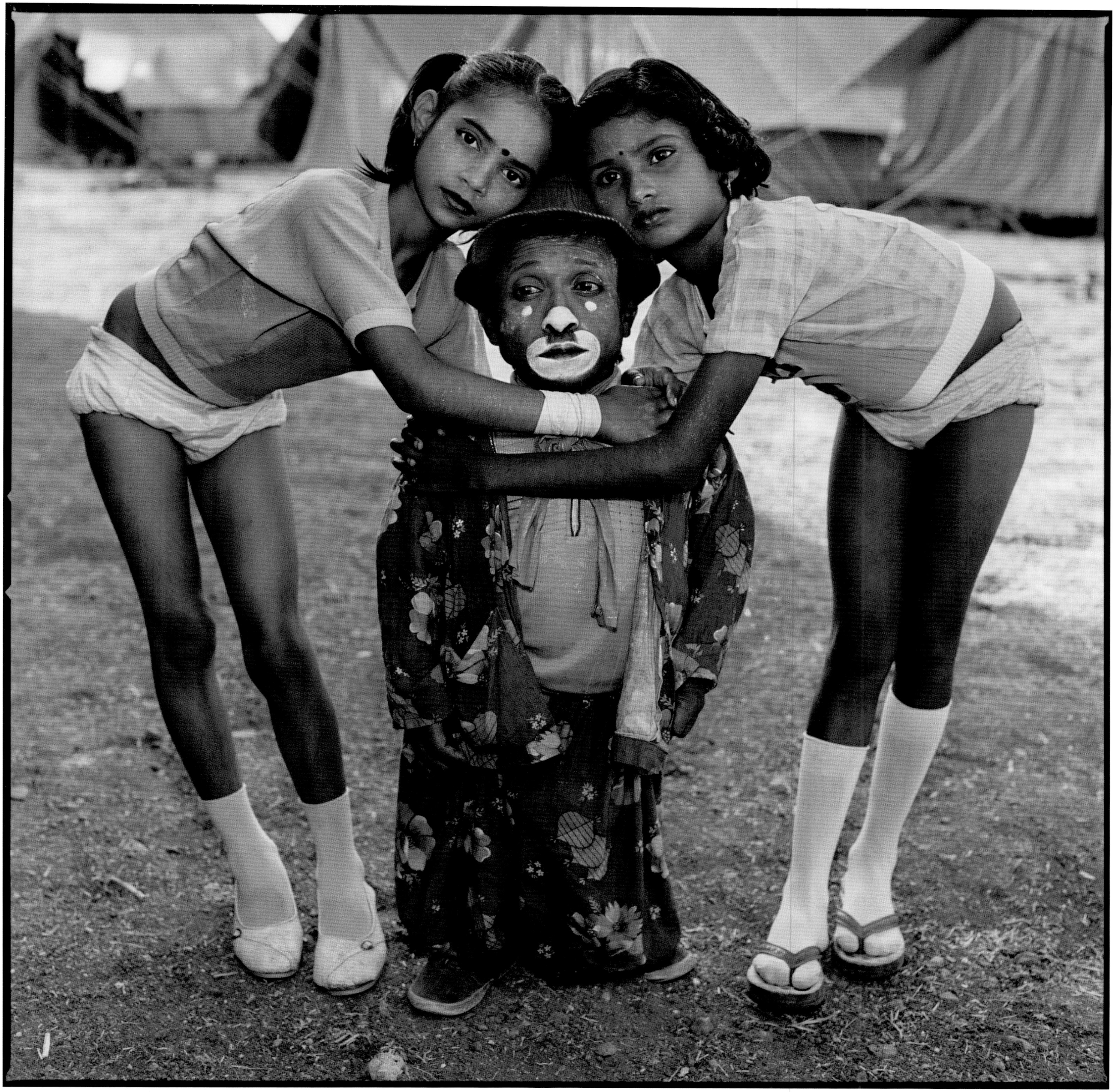

PLATE 109

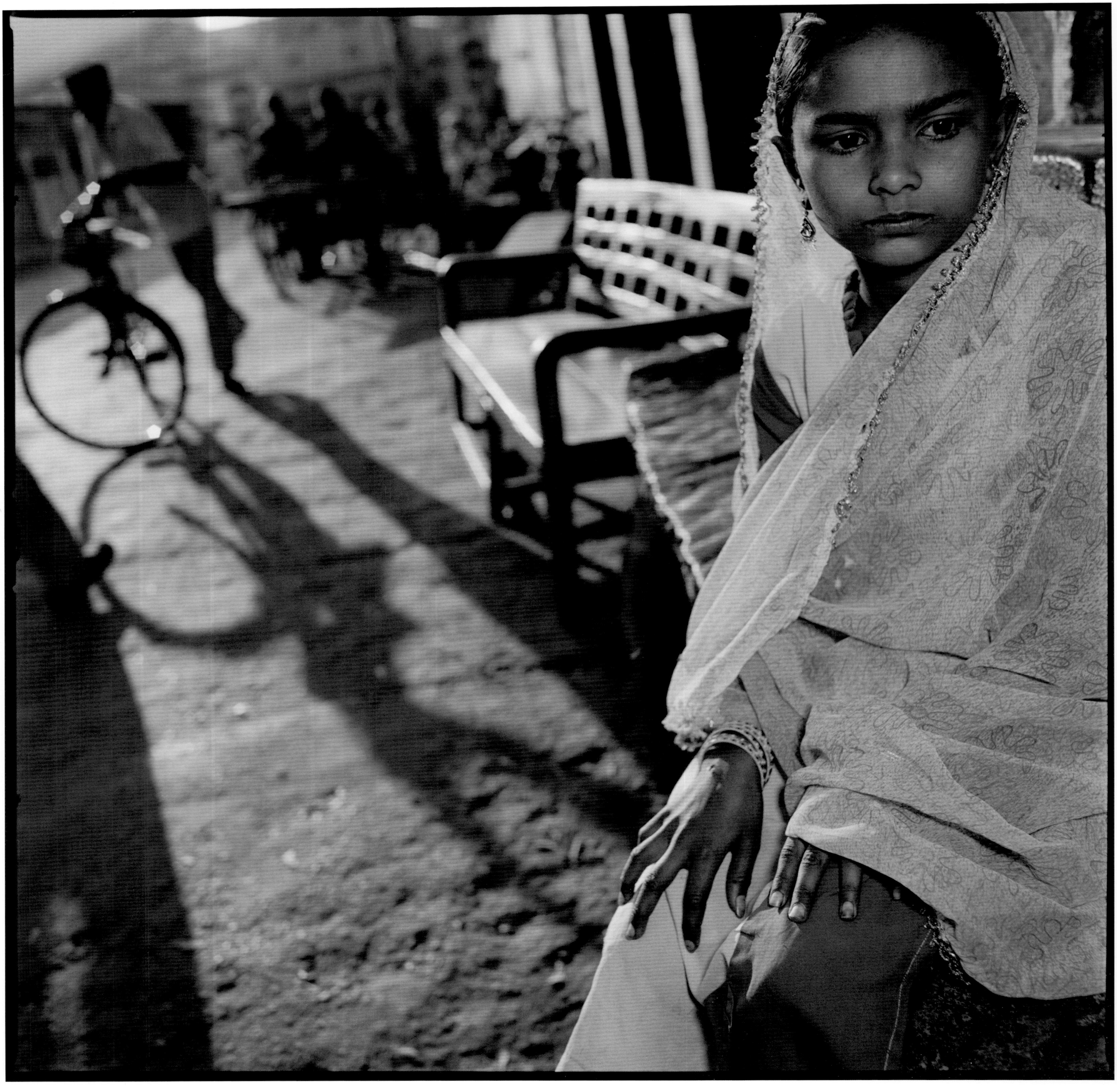

PLATE 110

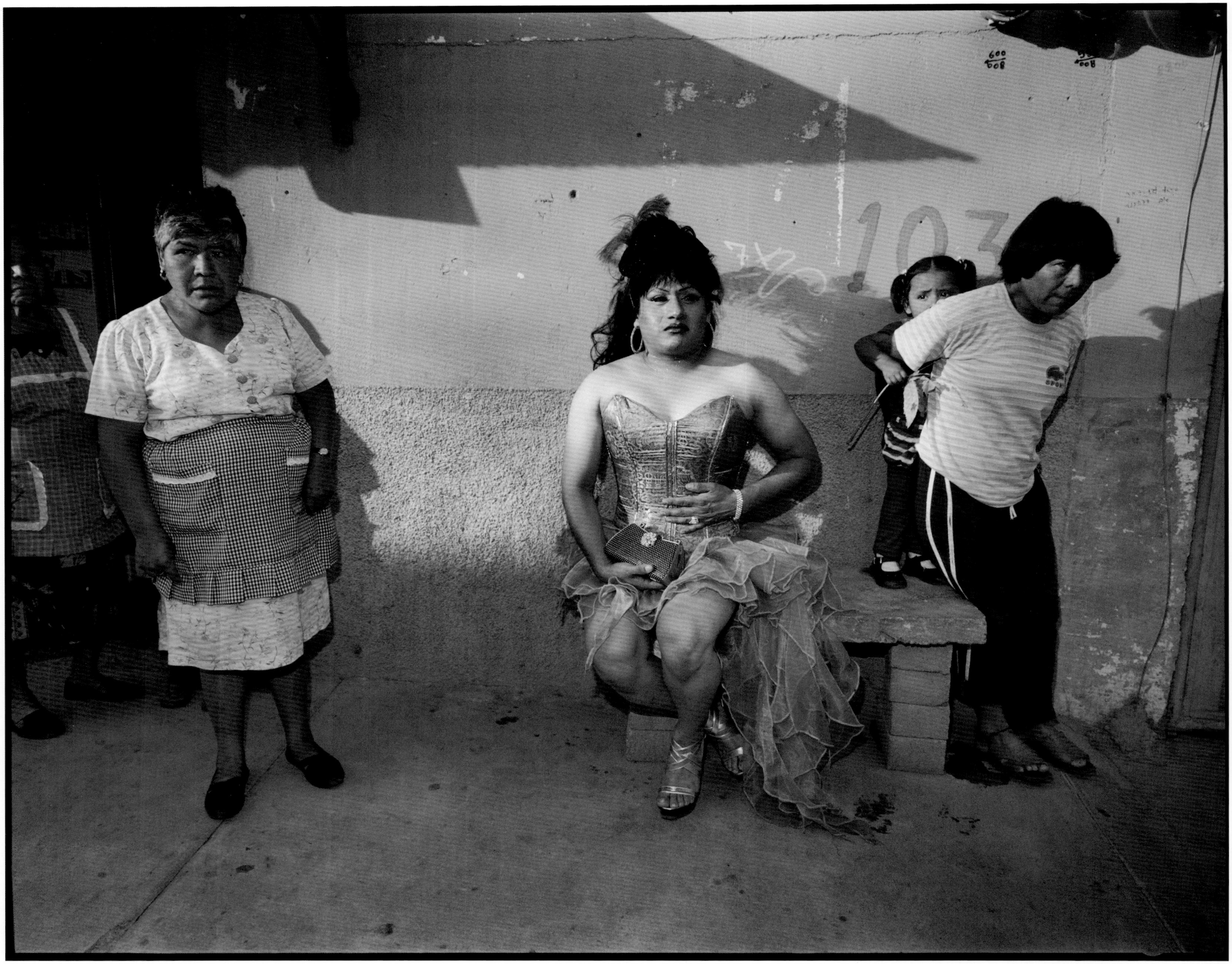

PLATE 111

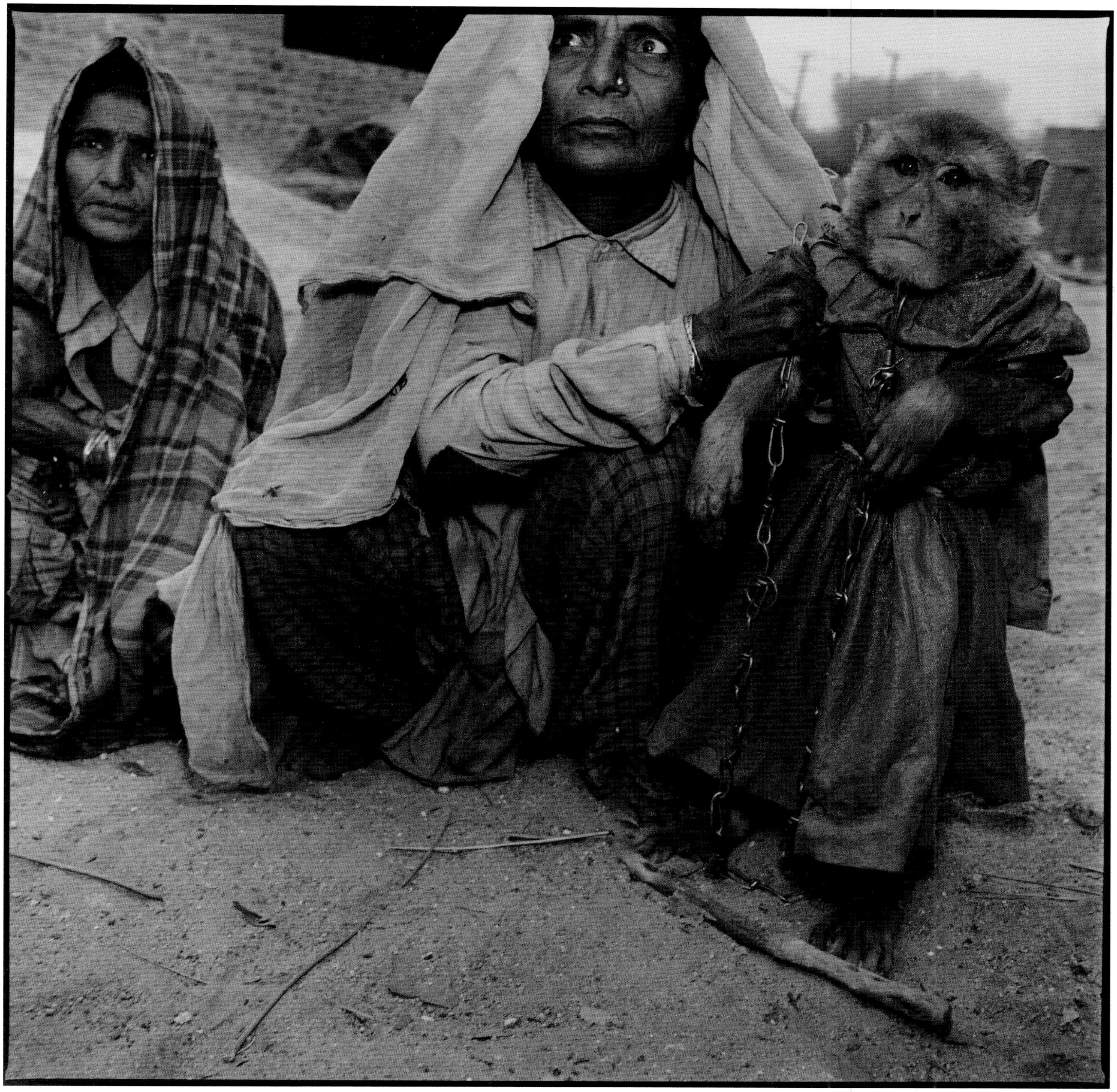

PLATE 112

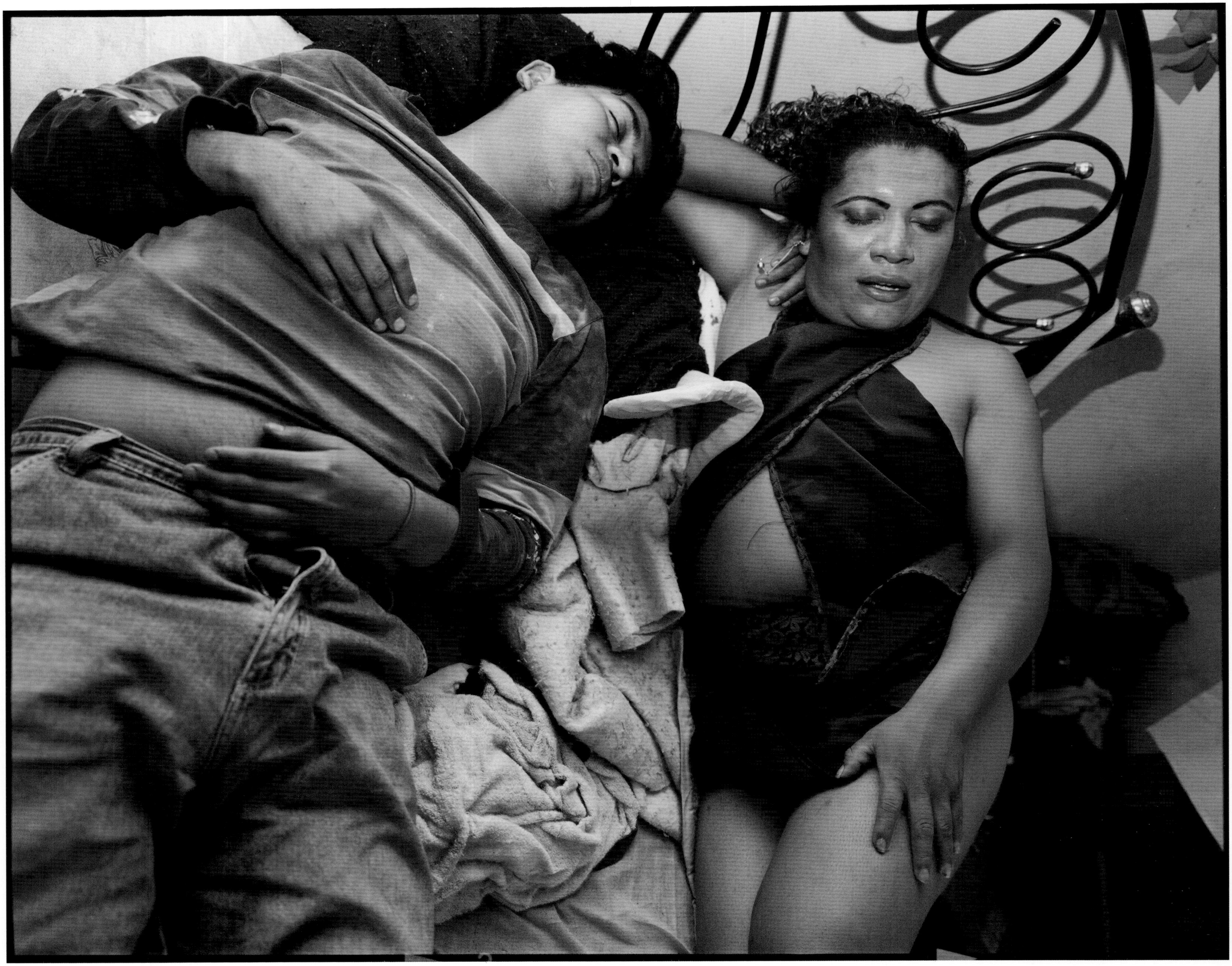

PLATE 113

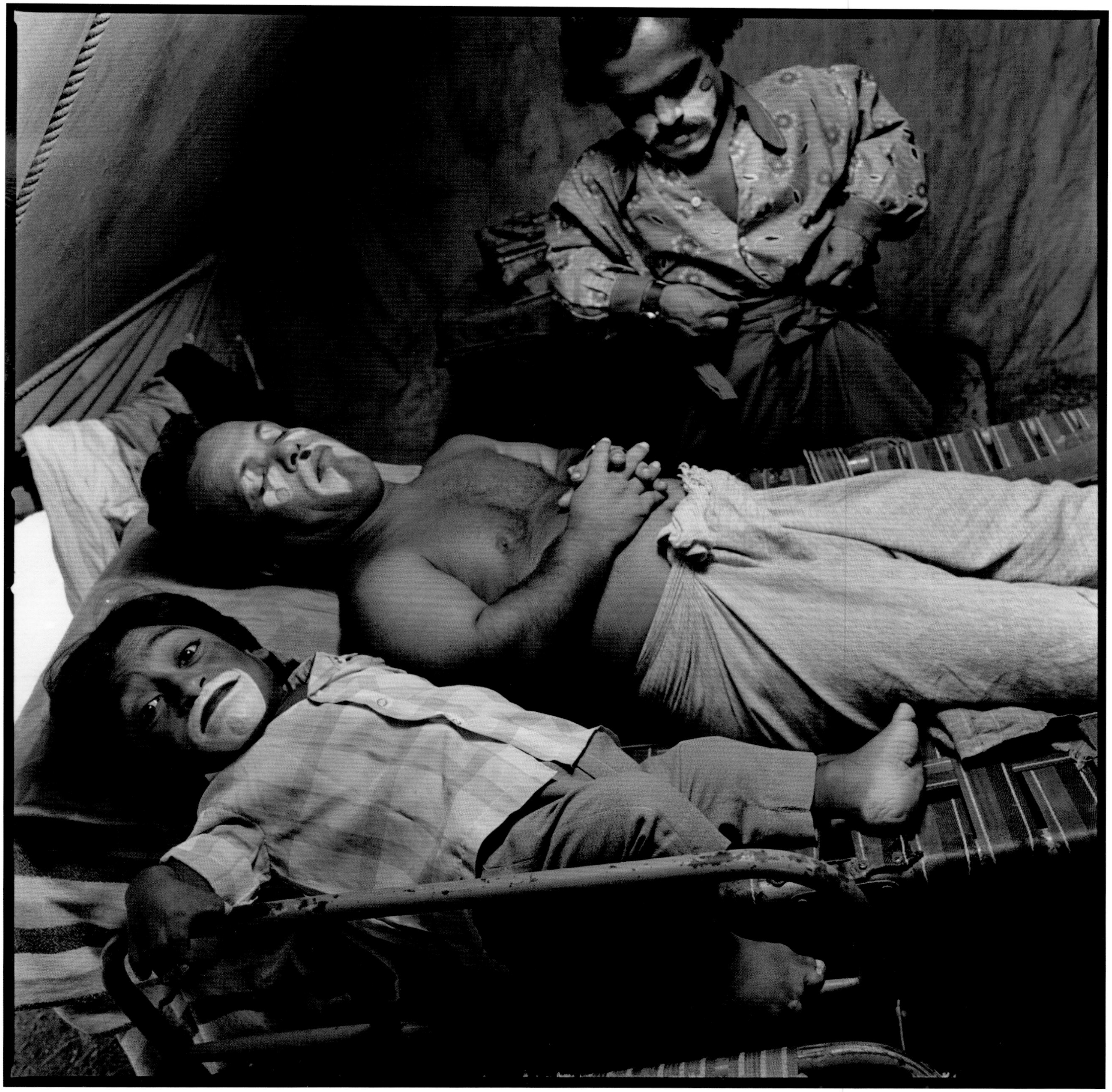

PLATE 114

PLATE 115

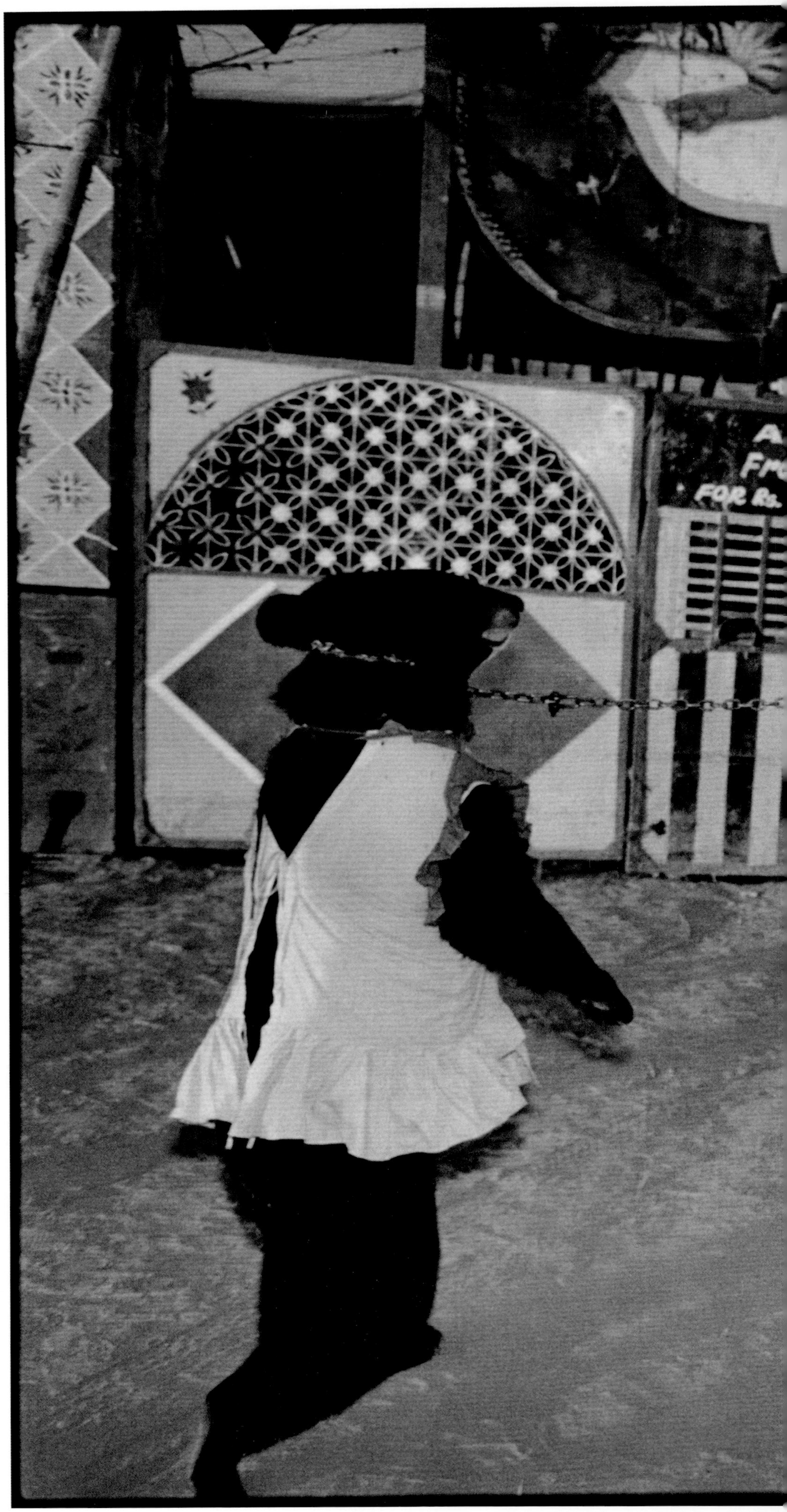

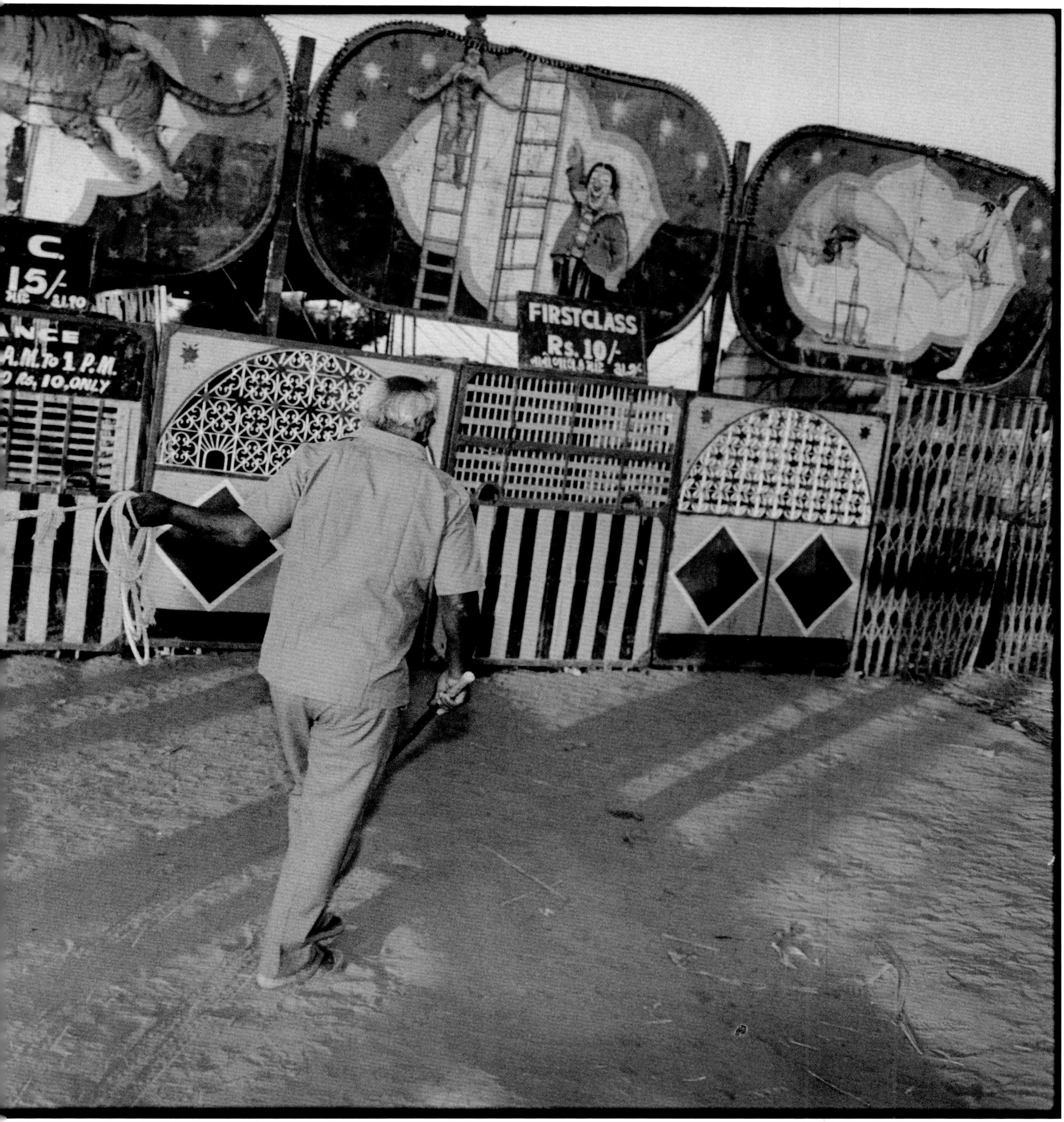
C.
15/-
A.M. To 1 P.M.
Rs, 10, ONLY
FIRSTCLASS
Rs. 10/-

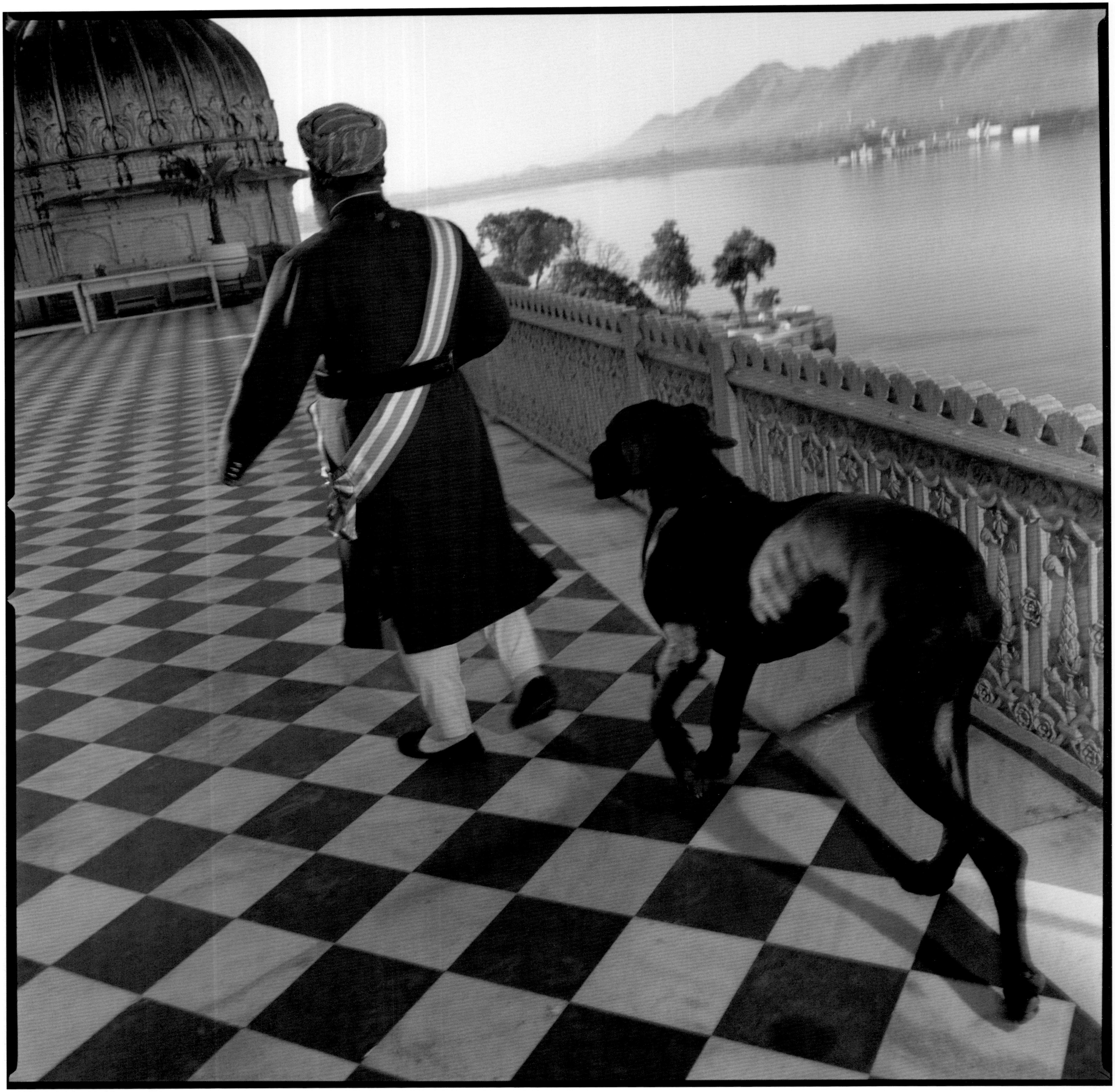

PLATE 116

PLATES LIST

PLATE 52 Monkey keeper with two monkeys on his shoulders, Raj Kamal Circus, Upleta, India, 1989.

PLATE 53 Three acrobats, Vázquez Brothers Circus, Mexico City, Mexico, 1997.

PLATE 54 Transvestite in backyard, Zaachila, Mexico, 2009.

PLATE 55 Festival, ladies and children in white, Oaxaca, Mexico, 1969.

PLATE 56 Mother Teresa praying on a train ride, Calcutta, India, 1980.

PLATE 57 Festival in "Old Goa," Mother Teresa posters, Goa, India, 2006.

PLATE 58 Mother Teresa's mission, Home for the Insane, Calcutta, India, 1980.

PLATE 59 Rolling to the church, Niño Fidencio, Espinazo, Mexico, 2000.

PLATE 60 People on the stairs, Oaxaca, Mexico, 1969.

PLATE 61 Sleeping dogs on the steps of the ghats, Benares, India, 1989.

PLATE 62 Iran, Circus d'Portugal, Mexico City, Mexico, 1997.

PLATE 63 Bald child mourning his father, Benares, India, 1989.

PLATE 64 Bathing girl, Niño Fidencio, Espinazo, Mexico, 2000.

PLATE 65 Jumping acrobats practicing, Famous Circus, Calcutta, India, 1989.

PLATE 66 Three young acrobats on a bed, Famous Circus, Calcutta, India, 1989.

PLATE 67 Carnaval masks with clown mask, Zaachila, Mexico, 2007.

PLATE 68 Acrobat with rings, Great Jumbo Circus, Mangalore, India, 1989.

PLATE 69 Transvestite and motorbike, Zaachila, Mexico, 2010.

PLATE 70 Monkey on a pony, Oaxaca Circus, Oaxaca, Mexico, 2005.

PLATE 71 Child acrobat with zebra, Oaxaca, Mexico, 2008.

PLATE 72 Marina Campa as Batman's grandmother, Kimberly Crown Circus, Mexico City, Mexico, 1997.

PLATE 73 Child acrobat with two children in peacock costumes, Great Royal Circus, Himmatnagar, India, 1989.

PLATE 74 Empire Circus, Calcutta, India, 1989.

PLATE 75 Veeru and Usman with Moti the Performing Vulture, Jumbo Circus, Mangalore, India, 1989.

PLATE 76 Child extra (with a saint) on *Old Gringo*, Mexico City, Mexico, 1988.

PLATE 77 Woman with snake, circus, Pune, India, 1992.

PLATE 78 Marching transvestites, Carnaval, Zaachila, Mexico, 2009.

PLATE 79 Ram Prakash Singh with his elephant Shyama, Great Golden Circus, Ahmedabad, India, 1990.

PLATE 80 Belén, tightrope walker, Garzetti Circus, Mexico, 1997.

PLATE 81 Children at the burning ghats, Benares, India, 1995.

PLATE 82 Four clowns, puppies, and a doll, Great Rayman Circus, Madras, India, 1989.

PLATE 83 Little girl clown with her Chihuahua, Oaxaca, Mexico, 2007.

PLATE 84 Dwarf with a rabbit and pigeon, Great Bombay Circus, Limbdi, India, 1990.

PLATE 85 Transvestite on a corner with a strange mask, Oaxaca, Mexico, 2009.

PLATE 86 Twin brothers Tulsi and Basant, Great Famous Circus, Calcutta, India, 1989.

PLATE 87 Clown shows his big fake feet in his hotel room, Oaxaca, Mexico, 2007.

PLATE 88 Two trainers and two chimps waiting to go on, Great Royal Circus, Himmatnagar, India, 1989.

PLATE 89 Gloria and Raja the chimp, Gemini Circus, Perintalmanna, India, 1989.

PLATE 90 Children in costumes with tourist behind, San Martín Tilcajete, Mexico, 2008.

PLATE 91 Dwarf with his wife, Amar Circus, Delhi, India, 1989.

PLATE 92 Daughters of "singing girls," Jodhpur, India, 1990.

PLATE 93 Transvestites waiting, Zaachila, Oaxaca, Mexico, 2010.

PLATE 94 Many child acrobats practicing, Great Golden Circus, Ahmedabad, India, 1990.

PLATE 95 Boy with a holy cow, Benares, India, 1995.

PLATE 96 Dwarf holding a rabbit, Bombay Circus, Ahmedabad, India, 1990.

PLATE 97 Pinky having her hair cut, Great Royal Circus, Cochin, India, 1992.

PLATE 98 Hippopotamus and performer, Great Rayman Circus, Madras, India, 1989.

PLATE 99 Madonna the giraffe, Oaxaca, Mexico, 1998.

PLATE 100 Dwarf clown with a balloon, Great Royal Circus, Goa, India, 1996.

PLATE 101 Usman holding his son, Jumbo Circus, Bombay, India, 1992.

PLATE 102 Hippo, Gemini Circus, Perintalmanna, India, 1989.

PLATE 103 Jyotsana riding on Vahini the elephant, Amar Circus, Delhi, India, 1989.

PLATE 104 Man dressed as a woman with a possum, Festival Parade, Zaachila, Mexico, 2007.

PLATE 105 Dwarfs line up, Jumbo Circus, Bombay, India, 1992.

PLATE 106 Children dressed in soldier costumes for a parade, Teotitlan, Mexico, 2010.

PLATE 107 Pinky practicing, Great Royal Circus, Cochin, India, 1992.

PLATE 108 Claudia and Maritza Safari with Babe the Performing Pig, Benneweis Circus, Mexico City, Mexico, 1997.

PLATE 109 Two girls and a dwarf, Raj Kamal Circus, Upleta India, 1989.

PLATE 110 Child on a bench, bicycle behind, Jodhpur, India, 1990.

PLATE 111 Transvestite by a shop, Zaachila, Mexico, 2010.

PLATE 112 Trainer with her monkey (monkey wearing a dress), Jodphur, India, 1990.

PLATE 113 Transvestites sleeping, Oaxaca, Mexico, 2006.

PLATE 114 Clowns sleeping, Jumbo Circus, Bombay, India, 1992.

PLATE 115 Runaway bear being led home, National Circus, Gujarat, India, 1989.

PLATE 116 Maharaja of Udaipur and his dog, Udaipur, India, 1996.

Thank you to everyone who made this book possible: Julia Bezgin, David Coleman, David Hamrick, Melissa Harris, Robert Hennessey, Chuck Kelton, Meredith Lue, Ellen McKie, and Bill Wittliff.

I want to thank Cyrus S. H. Jhabvala and Ruth Prawer Jhabvala for helping me to understand the peculiar beauty of India and therefore to love India forever.

Man and Beast
was designed by
Ellen McKie and Mary Ellen Mark

Text was set in Albertina

This book was printed on
157 gsm Chinese Gold East matte paper

Scans and color management by
Robert J. Hennessey Photography, LLC

Printed and bound by
Midas Printing International LTD